Bioinformatics in Agriculture

AF865034

NIPA® GENX ELECTRONIC RESOURCES & SOLUTIONS P. LTD.
New Delhi-110 034

Bioinformatics in Agriculture

Tools and Applications

M. Balakrishnan
Senior Scientists, (ICM Division)
National Academy of Agricultural Research Management
Rajendranagar, Hyderabad – 500 030
Andhra Pradesh

S. Dam Roy
Director (CARI)
Central Agricultural Research Institute
Port Blair – 744 101, Andaman & Nicobar Islands

NIPA® GENX ELECTRONIC RESOURCES & SOLUTIONS P. LTD.
New Delhi-110 034

NIPA® GENX ELECTRONIC RESOURCES & SOLUTIONS P. LTD.

101,103, Vikas Surya Plaza, CU Block
L.S.C. Market, Pitam Pura, New Delhi-110 034
Ph : +91 11 27341616, 27341717, 27341718
E-mail: newindiapublishingagency@gmail.com
www: www.nipabooks.com

For customer assistance, please contact
Phone: + 91-11-27 34 17 17
Fax: + 91-11- 27 34 16 16
E-Mail: feedbacks@nipabooks.com

ISBN: 978-81-19002-75-7

Composed and Designed by NIPA®.

राष्ट्रीय कृषि अनुसंधान प्रबंध अकादमी
राजेन्द्रनगर, हैदराबाद - 500 030, भारत
NATIONAL ACADEMY OF AGRICULTURAL RESEARCH MANAGEMENT
Rajendranagar, Hyderabad-500 030, INDIA
http://www.naarm.ernet.in
Phones : (040) 2401 5070; (R) : (040) 2401 5654; Fax : (040) 2401 5912; E-mail : director@naarm.ernet.in

डॉ. एस. एल. गोस्वामी, निदेशक
Dr S.L. GOSWAMI, Director

Foreword

Agriculture in the pre-green revolution era was relying heavily on the skills of the individual farmer. These skills were acquired through trial and error over generations and passed-on from one generation to another. The changes in socio-economic aspects and fragmentation of land holdings led to diverse occupations being adopted, with the result that the continuity of transfer of skill from one generation to the other was broken. Bioinformatics, one emerging mantras of modern agriculture has the potential to fill the gap in transfer of skills. Bioinformatics is an interdisciplinary area of the science composed of biology, mathematics and computer science. Bioinformatics is the application of information technology to manage biological data that helps in decoding plant and animal genomes. During the last two decades enormous data has been generated in biological science, firstly, with the onset of sequencing the genomes of model organisms and, secondly, rapid application of high throughput experimental techniques in laboratory research. Biological research that earlier used to be carried out in laboratories, fields and plant clinics, now carried out at computers (*In-silico*). Bioinformatics develops algorithms and data analysis tools to infer the information and make discoveries. Realizing the power of Bioinformatics, Indian Council of Agricultural Research, New Delhi, went in for computerization of biological data almost a decade and half back. In the initial stages, the requirements of hardware and connectivity issues were taken up along with appointment of manpower. Over a period of time, the Bioinformatics personnel got acquainted with the type of research activities carried out in the different institutes and also developed many decision support tools. I am happy that this book is contextualized around applications of Bioinformatics in Agriculture. It would give opportunity to the scientists to learn from each other and explore ways and means of augmenting the applications of Bioinformatics in agriculture and allied sectors.

The book contains various applications of bioinformatics in agriculture such as medicinal plants, corals, antibiotics, structural bioinformatics, text mining, data mining, protein structure prediction, drug design, gene expression, micro array, proteomics, molecular phylogenic location, rough sets to predict protein structural class, artificial neural network for prediction of amino acid levels, plant systems biology etc. This book also covers areas such as molecular modeling, homology modeling, bio-efficacy of micro-organism, fresh water aquaculture and fisheries, pesticide and insecticides, database and tool developments in the relevant area etc. It will be very much useful to the person working in agricultural biotechnology, Bioinformatics, Computer Science and Statistician etc.

I am sure that it will serve as a reference book for the development and application of Bioinformatics tools in agriculture.

Date
12.07.2013

S.L. Goswami

Preface

Agriculture is knowledge intensive and horticulture is more so, since Agricultural crops are generally high value, technology intensive crops. Bioinformatics offers the potential to help in the process of technology development as well as its delivery. Use of Bioinformatics tools in agriculture is recent in India. It is only just more than a decade that ICAR went in for development of Bioinformatics infrastructure and also appointed computer application personnel in its Institutes. Since then many initiatives in the field of Bioinformatics have been taken in the different Institutes and some of them are indeed laudable. However, for effective utilization of the hardware and human resources, there needs to be synergy between the subject specialists and Bioinformatics personnel. The subject specialist needs to know how to structure his knowledge so that Bioinformatics specialist can use it to develop decision support tools. The Bioinformatics specialist on the other hand needs to know what type of knowledge is available in different disciplines in agriculture and the type of information required by the different user groups so that decision support tools are designed appropriately to deliver the required information. In this respect we can learn from the experience of the developed countries where Bioinformatics initiatives met with limited success initially which led to the requirement of good decision support tools to be formulated. Therefore, we need to evaluate the Bioinformatics initiatives being undertaken in the different institutes in the light of the experience of developed countries and take corrective steps wherever needed.

This book is a step in the direction and is aimed at presenting the Bioinformatics initiatives of the different institutes working on Agricultural crops in the Indian Council of Agriculture Research. It is expected that this would help both the subject experts and Bioinformatics specialist to know the various Bioinformatics activities in agriculture and relate it to their own work. It is hoped that this book would be useful to the subject specialists to know how the information generated by him can be effectively utilized for problem solving. In many cases it would also help him to know what sort of information to generate so that his findings have greater practical utility. For the Bioinformatics specialists, this book would provide information on system design features of various tools developed in different Institutes so that he can adopt those suited to his own environment while developing similar tools. This book is expected to fill the gap of lack of sufficient literature in the field of Bioinformatics in agriculture in general and more pertinently to horticulture and that too for Indian conditions.

The Editors are grateful for the contributors of academicians, Scientist, research workers for their contribution for the book. We express our sincere thanks to the former Directors of CARI, Dr. R.C.Srivastava, and Dr. S.K. Ambast for their valuable suggestion and for taking the initiative to bring out this compilation in an excellent manner.

Editors

M. Balakrishnan
S. Dam Roy

Contributors

A.K. Roy
Central Institute of Freshwater Aquaculture
Kausalyaganga
Bhubaneshwar – 751 002
Odisha

A. Chatterjee
Department of Zoology
North Eastern Hill University
Shillong – 793 022, Meghalaya

Anusha Bhaskar
PG Department of Biotechnology
and Bioinformatics,
Bishop Heber College
Vyalur Road, Tiruchirappalli – 620 017

Arul Samraj D
Bioinformatics Centre of BTISnet
Madras Christian College (Autonomous)
Chennai – 600 059

Akash Solanki
Indira Gandhi Krishi Vishwavidyalaya
Raipur, Chhattisgarh – 492 006

A. Krithika
PG Department of Biotechnology
and Bioinformatics,
Bishop Heber College
Vyalur Road, Tiruchirappalli – 620 017

Amarendra Kumar Mishra
Indian Agriculture Research Institute
Pusa, Library Avenue
New Delhi – 110 012

A. Mahalakshmi
Lady Doak College
Madurai, Tamil Nadu

A. Subash Chandra Bose
J.J. College
Pudukkottai, Tamil Nadu

B.Varghese
Central Agricultural Research Institute (CARI)
Port Blair – 744 101
Andaman & Nicobar Islands

C.P. Rajadurai
AU-KBC Research Centre, MIT
Anna University Chennai, Chromepet
Chennai – 600 044

Christy Jancy Joy P
Bioinformatics Centre of BTISnet
Madras Christian College (Autonomous)
Chennai – 600 059

D. Velmurugan
CAS in Crystallography and Biophysics
University of Madras
Chennai

Dwijendra Gupta
Centre of Bioinformatics and Department of
Biochemistry
University of Allahabad, Allahabad
Uttar Pradesh

D.K. Biswal
Department of Zoology
North Eastern Hill University
Shillong, Meghalaya – 793 022

E. Hudson
PG Department of Biotechnology
and Bioinformatics
Bishop Heber College
Vyalur Road, Tiruchirappalli – 620 017

Grinson George
Fisheries Science Division
CARI, Port Blair, A&N Islands

G. Vidhya
PG Department of Biotechnology
and Bioinformatics
Bishop Heber College
Vyalur Road, Tiruchirappalli – 620 017

G. Ramesh Kumar
AU-KBC Research Centre, MIT
Anna University Chennai,
Chromepet Chennai – 600 044

K. Meena
Principal and Director
Shrimati Indira Gandhi College
Tiruchirappalli – 620 002
Tamil Nadu

K. Menaka
Department of I.T. & Applications
SIGC, Tiruchirappalli – 620 002
Tamil Nadu

K. Sarma
Scientist
Fisheries Science Division
CARI, Port Blair, A&N Islands

K. Balaji
PG Department of Biotechnology
and Bioinformatics
Bishop Heber College
Vyalur Road, Tiruchirappalli – 620 017

K.R. Subramanian
Lecturer (SS)
National College
Tiruchirappalli – 620 001
Tamil Nadu

K. Sangeetha
Bioinformatics Infrastructure Facility Centre of DBT
P.G. and Research Dept. of Microbiology and
Biotechnology
1&2 – BIFC of DBT
Presidency College (Aut.), Chennai – 600 005
Tamil Nadu

Kavita Singh
Central Agricultural Research Institute, CARI
Port Blair, A&N Islands

L. M. Goswami
Department of Zoology
North Eastern Hill University
Shillong, Meghalaya – 793 022

M. Ramachandran
Central Agricultural Research Institute, (CARI)
Port Blair, A&N Islands

M. Ramya
PG Department of Biotechnology
and Bioinformatics
Bishop Heber College
Vyalur Road, Tiruchirappalli – 620 017
Tamil Nadu

M. L. Lakhera
Indira Gandhi Krishi Vishwavidyalaya
Raipur, Chhattisgarh – 492 006

Mrs. Veena Tandon
Department of Zoology
North Eastern Hill University
Shillong, Meghalaya – 793 022

M. Balakrishnan
Senior Scientists, (ICM Division)
National Academy of Agricultural Research
Management
Rajendranagar, Hyderabad – 500 030
Andhra Pradesh

Mukunda Goswamy
4CSIR- Pool Scientist
NBFGR, Lucknow

Mayank Pokhriyal
Bioinformatics Centre, CARI
Port Blair – 744 101, A&N Islands

M. Rajadurai
PG Department of Biotechnology
and Bioinformatics
Bishop Heber College
Vyalur Road, Tiruchirappalli – 620 017
Tamil Nadu

Mrs. M. Manimekalai
Department of IT & Applications
Shrimati Indira Gandhi College
Trichy, Kerala

M. Durairaj
Shrimati Indira Gandhi College Tiruchirappalli
Tamil Nadu

M.Ramachandran
Central Agricultural Research Institute
ICAR, Port Blair – 744 101, A&N Islands

M.P. Singh
Indian Institute of Toxicology Research (IITR)
Lucknow – 226 001, U.P.

M. Yogambal
Bioinformatics Infrastructure Facility Centre of DBT
P.G. and Research Dept. of Microbiology and Biotechnology
1&2 – BIFC of DBT
Presidency College (Aut.), Chennai – 600 005

Michael Imanual Jesse
SRF,CARI
Port Blair – 744 101

M. Ramachandran
RA, CARI
Port Blair – 744 101

Navita Srivastava
Department of BioScience and Computer Science
Awadhesh Pratap Singh University
Rewa, Madhya Pradesh

Nibedita Jena
Central Institute of Freshwater Aquaculture
Kausalyaganga
Bhubaneswar – 751 002
Orissa

Priyakumari Joyce C
Bioinformatics Centre of BTI Snet
Madras Christian College (Autonomous)
Chennai – 600 059

P. Nithyanand
Department of Biotechnology Alagappa University
Karaikudi – 630 003, Tamil Nadu

P. K. Prasad
Department of Zoology
North Eastern Hill University
Shillong, Meghalaya – 793 022

Ravi R. Saxena
Indira Gandhi Krishi Vishwavidyalaya
Raipur, Chhattisgarh – 492 006

Ritu R. Saxena
Indira Gandhi Krishi Vishwavidyalaya
Raipur, Chhattisgarh – 492 006

R. Shenbagarathai
Lady Doak College
Madurai
Tamil Nadu

R.W. Alexander Jesudasan
Department of Zoology
Madras Christian College (Autonomous) Tambaram East, Chennai – 600 0509

R.C. Srivastava
Central Agricultural Research Institute, CARI
Port Blair – 744 101
Andaman & Nicobar Islandss

Rajni Jain
National Centre for Agricultural Economics and Policy Research
Pusa, IARI Campus
New Delhi – 110012

R. Venkateshwaran
Maruthupandiyar College
Thanjavur, Tamil Nadu

S.N. Sethi
Madras Research Centre of CMFRI
ICAR
75, Santhom High Road
R.A. Puram, Chennai – 600 028

S. Gifta Lorraine
Bioinformatics Centre of BTISnet
Madras Christian College (Autonomous)
Chennai – 600 059

S.R. Martha
Central Institute of Freshwater Aquaculture
Kausalyaganga
Bhubaneshwar – 751 002
Odisha

S. Dom Roy
Fisheries Science Division
Central Agricultural Research Institute, (CARI)
Port Blair – 744 101, A&N Islands

S. Jeyakumar
Animal Science
CARI, Port Blair – 744 101, A&N Islands

Soma Mondal
Senior Research Fellow
Bioinformatics Centre
Central Agricultural Research Institute, (CARI)
Port Blair – 744 101, A&N Islands

S. K. Zamir Ahmed
Senior Scientist, CARI
Port Blair- 744 101, A&N Islands

Subhash Chand
Senior Scientist, CARI
Port Blair – 744 101, A&N Islands

S. Manju
Department of Biotechnology
Alagappa University
Karaikudi – 630 003, Tamil Nadu

S. Karutha Pandian
Department of Biotechnology
Alagappa University
Karaikudi – 630 003, Tamil Nadu

S. Rajarajan
Bioinformatics Infrastructure Facility Centre of DBT
P.G. and Research Department of Microbiology
and Biotechnology
1&2 – BIFC of DBT
Presidency College (Aut.)
Chennai – 600 005

T.V. Sundar
Department of M.C.A.
Shrimati Indira Gandhi College
Tiruchirappalli – 620 002
Tamil Nadu

Vidya Murali
PG Department of Biotechnology
and Bioinformatics,
Bishop Heber College
Vyalur Road, Tiruchirappalli – 620 017
Tamil Nadu

V. Stanley
PG Department of Biotechnology
and Bioinformatics,
Bishop Heber College
Vyalur Road, Tiruchirappalli – 620 017
Tamil Nadu

Vijay Tripathi
Centre of Bioinformatics and
Department of Biochemistry
University of Allahabad, Allahabad
Uttar Pradesh

V. Tandon
Department of Zoology
North Eastern Hill University
Shillong, Meghalaya – 793 022

Contents

1

Application of Bioinformatics in Agriculture

D. Velmurugan

Bioinformatics is the application of information technology to the field of molecular biology. The term Bioinformatics was coined by Paulien Hogewag, in 1978 for the study of informatic processes in biotic systems. Bioinformatics nowadays, entails the creation and advancement of databases, algorithms, computational and statistical techniques, and theory to solve formal and practical problems arising from the management and analysis of biological data. Bioinformatics deals with storing and searching of large biological data sets and analysis of these data sets using computers to determine various relationships.

The rapid developments in genomic and other molecular research technologies and developments in information technologies have been combined to produce tremendous amount of information related to molecular biologists. Now bioinformatics has developed and deals with many other types of biological data like protein structures, sequence alignment, gene finding, genome assembly, gene expression profiles, structure alignment, protein structure prediction, protein–protein interactions and the modelling of evolution. Being an interface between modern biology and informatics, it involves detection, expansion and implementation of computational algorithms and some of the software tools used to provide applications primarily in agriculture and healthcare divisions. In developing countries, bioinformatics play a major role in agricultural area where it can be used for increasing the nutritional content, and increasing the volume of the agricultural products *etc.*

BIOINFORMATICS IN AGRICULTURE

Agri Informatics is based on a GIS (Geographical Information System) analysis of the complex interaction between the environment and crops. Analysis of the micro climate can be draw by the use of simple maps of the farm. According to a variety of critical variables, the ideal block borders are drawn. In the same way, a soil map is also drawn up for describing soil qualities such as the acidity level, clay content, water retention ability and depth. The result of the map shows, the ideal block design, row alignment, root stock, cultivator, plant spacing, and trellis system.

The farm meso-climate is compared to other wine region, locally and internationally. These analyses form the basis of precision viticulture. Aerial photography or satellite images can also be included in the system. Satellite based remote sensing technique combined with limited field survey provide valuable information in rapid time. The information provided by the data which has been generated on the status of crops under observation at any particular date can be processed quickly. This advantage provides a synoptic view, making it possible to analyse the status and trend of transplantation, crop growth and harvesting throughout the study area. Application of remote sensing and GIS will be used for creating a systematic and sharable database on crop related subject and answer the unanswered questions of different stakeholders.

EVOLUTIONARY STUDIES

To determine the tree of life and the last universal common ancestor, the sequencing of genomes from all three domains of life, eukaryota, bacteria and archaea means that evolutionary studies can be performed in a mission.

CROP IMPROVEMENT

Comparative genetics of the plant genomes has shown that the organization of their genes is more conserved over evolutionary time than was previously believed. These results suggest that information obtained from the model crop systems can be used to suggest improvements to other food crops. At hand the complete genomes of Arabidopsis thaliana (water cress) and Oryza sativa (rice) are available.

INSECT RESISTANCE

Bacillus thuringiensis genes control a number of serious pests that have been successfully transferred to cotton, maize and potatoes. This new aptitude of the plants to resist insect attack means that the amount of insecticides being used can be reduced and hence the nutritional quality of the crops is increased.

IMPROVE NUTRITIONAL QUALITY

The recent development in agriculture is transferring genes into rice to increase levels of Vitamin A, iron and other micronutrients. This work might have a deep impact in reducing occurrences of blindness and anaemia which is caused by deficiencies in Vitamin A and iron respectively. Another one is, to insert a yeast gene into the tomato, and the result is a plant whose fruit stays longer on the vine and has an extended shelf life.

DEVELOPMENT OF DROUGHT RESISTANCE VARIETIES

Development has been made in developing cereal varieties that have a greater tolerance for soil alkalinity, free aluminium and iron toxicities. These types of varieties will allow agriculture to succeed in poorer soil areas, thus adding more land to the global production base. The development work is in progress on the production of crop varieties capable of tolerating reduced water conditions.

IN THE FIELD OF GENETIC ENGINEERING

Plant Transformation

Genetic engineering is also using the plants as factories for pharmaceutical protein production. By inserting a segment of DNA that determines the protein of choice into plant cells, the plant-made pharmaceuticals are made.The plants or plant cells are important factories which are used to produce the desired proteins and are only grown for the purpose of pharmaceutical applications.

Transformation containing two common methods (the process by which DNA from one organism is incorporated into the DNA of another organism) have been recognized through biotechnology to produce transgenic plants, which, in turn, could be used to create the plants used to make pharmaceutical proteins. The transformation techniques include the *Agrobacterium tumefaciens*-mediated transformation system and biolistics, also called particle bombardment.

Agrobacterium tumefaciens is a bacterium that naturally infects plants and causes crown gall disease. It is used for the production of transgenic plants because it has the amazing ability to transfer a segment of its DNA, called T-DNA, into the nucleus of the plant cells. The T-DNA from *A.tumefaciens* is then incorporated into the plant and transcribed, causing crown gall disease. *A.tumefaciens*is used for the DNA of interest between the T-DNA to create a plant with qualities such as herbicide resistance or pesticide resistance, as well as many others (Nester et al. 1984, Binns and Thomashow 1988). Crown gall disease does not take place in the plant after being transformed by *A.tumefaciens* because the plasmid has been constructed to disarm the tumor-inducing properties which are cause the disease, and as a substitute it expresses the DNA of interest.

The particle bombardment transformation system employs the use of metal particles, either of gold or tungsten. The metal particle is to be transferred into the plant cells. Using a pressurized system, a plastic bullet covered with metal and DNA particles is released from the biolistics machine when the pressure is released.

Within seconds the bullet shot is stopped by a shield causing the metal coated DNA particles to be knocked off the bullet. The particles are then ultimately forced to be inserted into the plant cells below.

The cells expressing the gene of interest must be selected when the genes have been transferred into the cells. In some cases, a selectable marker is included with the DNA construct, which will confer resistance to an antibiotic or herbicide. The plant cells that survive the antibiotic or herbicide application are the ones that have successfully received the gene of interest (Purdue Agricultural Biotechnology, 2004).

PLANT SYSTEMS *Vs.* OTHER CELL CULTURES

Plants and plant cells offer many advantages in the production of proteins for pharmaceutical purposes. Different mammalian cells, bacterial cells and plants are accomplished of post-translational modification and other assembly steps that are needed for biological activity in complex multi-component proteins such as antibodies. This in turn allows for more diversity and possibilities of drugs that could be imagined from the proteins produced in plants (Rogers 2003; Goldstein and Thomas, 2004). A certain sugar molecule is added to its structure when a protein is glycosylated causing it to fold in a specific and unique orientation. The way the protein is folded determines its function and activity. Plants are able of producing more complex proteins with no assistance because they are capable of glycosylation on their own, whereas mammalian cells need assistance to form glycosylate complex proteins. The system needed to preserve and make complex proteins in mammalian cells is expensive and difficult to maintain, which makes plant systems a more attractive source for protein production. (Fischer *et al.*, 2003).

Mammalian cells are difficult and expensive to grow; therefore, bacterial cells are used to produce proteins when glycosylation is not required. Bacterial cells have the potential of being contaminated, and require facilities to build to keep up with the growing demand of proteins needed for drug production. So, plants are excellent alternatives to mammalian and bacterial cell cultures and they have the ability to glycosylate complex proteins and are inexpensive to grow in large quantities.

PLANTS IN PHARMACEUTICAL FIELD

Plant-made pharmaceutical production has the possible to provide large amounts of proteins for the pharmaceutical industry. Different types of plants used for pharmaceutical protein production depend upon their final applications. To date, the plants that have pharmaceutical applications include tobacco, potato, tomato, corn, soybeans, alfalfa, rice, and wheat. However, it is important to emphasize that no commercially available pharmaceutical products are currently produced in plants (Ma *et al.,* 2003).

Tobacco was the first plant to be genetically engineered and has the advantage of being used as a plant biopharmaceutical for exactly that reason; the methods for gene transfer and expression are well recognized for this plant. It also can be cropped multiple times per year, giving it an edge over other plants in producing a large biomass (Fischer *et al.,* 2003). The protein of interest can also be targeted to the seed. Up to one million seeds can be made on a single tobacco plant (Daniell *et al.,* 2001). One concern associated with using tobacco as a protein factory is that it has toxic alkoloids associated with it, which could be present in the final drug that is administrated to the patient. Another concern is that the protein must be frozen or dried and then development immediately, if the protein is targeted for the leaf tissue, making it hard to harvest and store for preservation of the protein (Ma *et al.,* 2003).

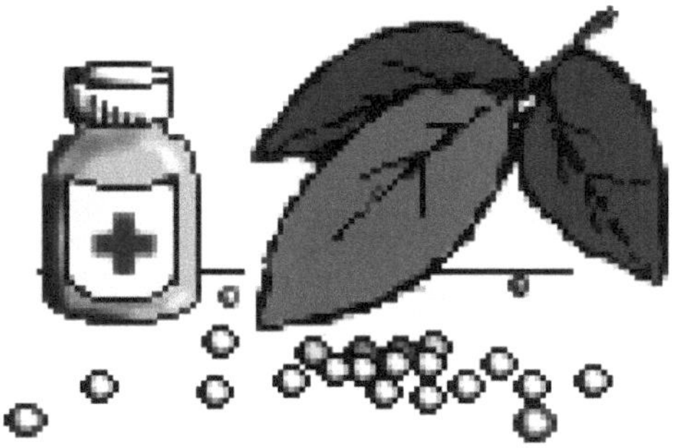

APPLICATION

The application of the proteins made in plants include, antibiotics, vaccines, hormones, enzymes, interleukins, interferons, and human serum albumins. Antibodies are produced in vertebrate immune systems to recognize and bind to antigens with amazing specificity. There are six types of plant-derived recombinant antibodies that have moved to the preclinical stage of testing, and one more advanced antibody in phase II clinical trial (Ma *et al.,* 2003).

Another application for plant-made pharmaceuticals is the production of vaccines. Subunit vaccines are prepared by the use of specific macromolecules that induce a protective immune response against a pathogen. Subunit vaccine technology has increased the safety of administrating vaccines because it does not involve the use of live or weakened viruses.

POSSIBLE RISKS

Potential risks associated with plant-based pharmaceuticals include, pollen transfer to related species, contamination of non-transgenic crops intended for the use by humans, allergic reactions to the drugs produced from the genetically engineered plant, and determination of genetically engineered material to persist in the environment and accumulate in non-target organisms (Daniell *et al.,* 2001). Risk assessment of plant-made pharmaceuticals should be reviewed on a case-by-case basis because the plants used to produced proteins each have different risks associated with them (Peterson and Arntzen, 2004).To address the out-crossing potential of transgenic crops to a receptive crop species or weedy relative, physical isolation distances have been considered. Some species have enclosed flower structures, reducing the distance needed in between fields and therefore are being evaluated as good candidates for producing plant-made pharmaceuticals. In the case of corn, a wind pollinated crop because it has many benefits for being a plant biopharmaceutical, the tassels can be manually removed and male-sterile varieties could be used to contain the pollen. The transgenic plants can also be planted at different times from food crops to make sure they flower at different times, which decreases the potential for pollen transfer (Thomas *et al.,* 2002).

IN BIO-FUELS

With the point in oil prices in 2005, much attention has turned to the utilization of alternative fuels. A popular alternative fuel source is bio-fuel. It is a broad term that encompasses both bio-diesel and ethanol. It is a fuel not derived from fossils, but is instead derived from one of two sources: agricultural crops such as corn, soybean, and sugar cane, or biomass resources such as agricultural, wood, animal, and municipal wastes and residues.

Bio-fuel is a renewable source of fuel used as liquid fuels for transportation, but it also can be used for direct combustion for electricity production. Bio-diesel is different from ethanol because it is made from the oils of plants, mainly soybean, sunflower and waste vegetable oil. It contains no petroleum, is free of sulphur and aromatics, and can be blended at any level with petroleum diesel, usually called bio-diesel blend.

ETHANOL PRODUCTION

The chemical process of turning cellulose and starch into dimethyl ether (ethonal) first requires breaking the chemicals into single sugar molecules. In plant both cellulose and starch are a polysaccharide or string of many individual sugar molecules attached together. When sugars are broken down, these are fermented into ethanol. This needs to be distilled before it is ready for automobiles (Moreira 2005). An organism will break down the cellulose and ferment the sugars in one step. This will reduce the cost of producing ethanol and make the process feasible for quicker scale-up. After the ethanol is produced, it is merged with petroleum-based fuel that can be pumped directly into the gas tank of our current vehicle models depending on the ethanol/petroleum merge.

All cars can run on E10, which is 10 percent ethanol and 90 percent petroleum, but cars need to be altered to run efficiently on a blend any higher than E10 (for example E20, E85). Cars cannot tolerate high concentrations of ethanol because ethanol has higher oxygen content than petroleum-based fuel and cars need to be able to detect this with a fuel sensor. Other car parts exposed to ethanol need to be modified because it is an alcohol and it can disrupt the parts.

ENVIRONMENTAL BENEFITS

The popularity of bio-fuels is increasing because of the many benefits, especially in a world that is becoming more environmentally conscious and pressured by the instability of oil supplies and prices. For the production of bio-fuel, crops are used and take in carbon dioxide from the atmosphere and some argue that this offsets the greenhouse gases released when the fuel is burned.

Replacing petroleum with bio-fuel has the potential of reducing air pollution, including emissions of fine particulates and carbon monoxide. Greenhouse gas emissions from vehicles are reduced when vehicles run on any blend of 10% ethanol and can potentially reduce greenhouse gas emissions by 10 grams of carbon dioxide for every kilometre driven (Iogen Corporation, 2006).

Another benefit is that rural economics may have the chance to maintain themselves and possibly grow because the process of making bio-fuels creates jobs and many raise farm incomes. It can be produced locally and at the same time is a renewable fuel, giving opportunities to diversify energy sources and lower the dependence on foreign oil. If the bio-fuels become a main energy source, the benefits are plenty, but it will only make the biggest difference in the environment. The largest obstacle to this is land availability, so land efficiency or maximizing energy yield per acre is an essential aspect when considering bio-fuels as a main energy source. The impact of biotechnology is making this latter point less of a challenge with advances in crop yield productivity.

When considering the input/output ratios of energy, plant efficiency is important. For example, corn is one of the least efficient sources of ethanol. Compared to corn, sugar beets and sugarcane produce double the ethanol yield per acre. The energy input to make ethanol from corn (growing, transporting, and distilling) uses almost as much energy contained in the ethanol itself. Sugarcane yields eight times much energy that is needed to produce ethanol compared to corn. Unfortunately, with growing fuel demands in tropical countries like Brazil, fuel demand translates into increased sugarcane production, meaning a demand for more fields, resulting in reductions in rainforest area (Murray 2005).

IN GENOMICS

Plant Genomics

Plant genomics is the field of study that considers plant genomes: the sequence, arrangement of genes and other elements on chromosomes, and their evolution. The plant cell contains three different genomes like, the chloroplast, the mitochondrian and the nucleus. Data from all the three genomes have been used under different conditions to get better considerate of the diversity and distribution of plants.

The chloroplast and mitochondrian are usually inherited maternally in angiosperms and the nucleus is biparental. They differ dramatically in size, with the nucleus being by far the largest measured in megabases. The mitochondrial genome includes several hundred kilobase pairs (kbp) of DNA (200-2500 kbp), which is comparatively smaller to the nuclear genome. In most of the plant, the chloroplast genome is the smallest of the three plant genomes (135-160 kbp). In the mitochondrian, the order of the genes is variable and they are separated by the large regions of non-coding DNA. The mitochondrial genome rearranges itself frequently, so that many rearranged forms can occur in the same cell within individual plants that they do not characterize or differentiate species or groups of species. Consequently, they are not especially useful for inferring relationships. The chloroplast, in contrast, is stable, both within the cells and the species. The presence of two regions that encode the same genes, is the clearest feature, but in opposite directions known as inverted repeats. A small single copy region and a large single copy region are found in between them. Rearrangements of the chloroplast genome are rare enough in evolution that they can be used to demarcate major groups. The order of genes in the nuclear genome is also presumed to be stable, at least within species, and may be stable across groups of species as well.

Type	Genome size (kbp)	Inheritance
Chloroplast	135-160	Generally maternal (from the seed parent)
Mitochondrian	200-2500	Generally maternal (from the seed parent)
Nucleus	1.1-110	Biparental

AGRICULTURAL RESEARCH SERVICE AND NATIONAL RESEARCH INITIATIVE

The Agricultural Research Service (ARS) is concerned with mapping databases through a rather simple but efficient process. Funding, directly and indirectly, is supplied by ARS to "database curators" for the various species. These researchers take the lead for their respective species, perform some evaluation and quality control of their databases and hold that information in the home laboratories of their institution. The information is sent to the USDA National Agricultural Library, where it is integrated into a master or central database system with data from all of the species.

OBJECTIVES OF THE PLANT GENOME PROGRAMME

1. This program is one of the part of the US department of agriculture plant genome research program, the goals are to foster and coordinate research to identify, characterize, alter and rapidly and precisely manipulate genes that control plant traits important to the productivity and sustainability of agriculture and forestry.
2. Development of new technologies for genome mapping, genome manipulation, gene isolation, or gene transfer in plants.
3. Application of genome technologies to generate new products important to U.S. agricultural producers and consumers.
4. Progress of plant genome software to support Comparative mapping, Quantitative trait loci (QTL) analysis, and Pedigree analysis.

PACIFIC DIGITAL LIBRARY EXPERIMENT PLANT GENOME DATABASES

The focus is the collection, organization, integration, and delivery of genome information for agriculturally important organisms. The plant genome data come primarily from curators who are responsible for collecting, organizing and evaluating data for individual taxa. The resulting data is then forwarded to the National Agricultural Library where it is made available to users via the www server as well as in other forms. Curators maintain data resources as their home institutions. These resources may include online databases, data that have not been entered into a database, collections of images, tarred and compressed versions of databases themselves. The potential benefit is the sharing of plant genome data which is critical to good scientific research.

PLANT GENOME DATABASES

Plant genome database is an information resource being developed at NASC (Nottingham Arabidopsis Stock Centre) as part of the UK – crop net project. The main aim is to integrate data with physical and genetic maps of plant genome, within the UK crop net project and provide information for identifying gene function and crop plant orthologues. The different types of plant genome databases are available in their respective database Home pages:

1. Arabidopsis thaliana database – Aa TDB
2. Sorghum database
3. Mendel database
4. Cotton database
5. Grain gene database
6. Millet gene database
7. Rice gene database
8. Pathogenes database

BT BRINJAL

Brinjal or baigan is also known as eggplant and aubergine in North America and Europe respectively. It is a very important comman man's vegetable in India. It is often illustrated as a poor man's vegetable because it is a popular amongst small-scale farmers and low income consumers.

INDIA'S FIRST VEGETABLE BIOTECH CROP

FSB-resistant brinjal or Bt brinjal was developed using a transformation process in the development of Bt cotton, a biotech crop that was planted on 6.2 million hectares in India in 2007. Bt brinjal incorporates the *cry1Ac*gene expressing insecticidal protein to confer resistance against FSB. The *cry1Ac* gene is taken from the soil bacterium *Bacillus thuringiensis* (Bt). When ingested by the FSB larvae, the Bt protein is activated in the insect's alkaline gut and binds to the gut wall, which breaks down, allowing the Bt spores to invade the insect's body cavity. The FSB larvae die a few days later.

Bt Brinjal was developed by the Maharashtra Hybrid Seeds Company (Mahyco). To transform young cotyledons of brinjal plants, the company used a DNA construct containing the *cry1Ac* gene, a CaMV 35S promoter and the selectable marker genes *nptII* and *aad*. A single copy elite event, named EE-1, was selected and introduced into hybrid brinjal in Mahyco's breeding program. Mahyco also donated the technology to public research institutions in the Philippines and Bangladesh. Several other research institutions, both public and private have also

been developing Bt brinjal using different genes. The NationalResearchCenter on Plant Biotechnology (NRCPB) has developed Bt brinjal varieties expressing the *cryFa1* gene. The technology was subsequently transferred to Pocket K No. 35 Bt Brinjal in India http://www.isaaa.org/kc. Using the *cry1Ab* gene the Indian Institute of Horticultural Research (IIHR) is also developing Bt brinjal. Scientists are also looking for ways to develop Bt brinjal in conjunction with other multiple and beneficial traits.

BIOSAFETY AND FOOD SAFETY ASSESSMENTS

Exact scientific tests, including toxicity and allergenicity evaluation as well as nutritional studies on rabbits, rats, carps, goats, broiler chickens and dairy cows, have confirmed that Bt brinjal is as safe as its non-Bt counterparts. The results of the studies on pollen escape, effects on soil microflora and non-target organisms, agronomy, invasiveness and Bt protein degradation are validated for the safety of Bt brinjal. Results of the studies demonstrated that Bt brinjal does not affect beneficial insects such as aphids, leafhoppers, spiders and lady beetles.

REFERENCES

Choudhary, B and Gaur, K. 2008. The Development and Regulation of Bt Brinjal in India (Eggplant/Aubergine), *ISAAA Brief* No. 38, ISAAA: Ithaca, NY.

Daniell, H., Streatfield, S. J., Wycoff, K. 2001. Medical molecular farming: production of antibodies, biopharmaceuticals and edible vaccines in plants. *Trends in Plant Science* 6:219-226.

Fischer, R., Twyman, R. M., Schillberg, S. 2003. Production of antibodies in plants and their use for global health. *Vaccine* 21:820-825.

James, C. 2007. Global Status of Commercialized Biotech/GM Crops 2007. ISAAA Brief No.37. ISAAA. Ithaca, New York, USA.

Krishna, V.V. and Qaim, M. 2007. Estimating the Adoption of Bt Eggplant in India: Who Benefits from Public-Private Partnership?, *Food Policy*, pp. 523-543.

Ma, J.K-C., Drake, P.M.W., Christou, P. 2003. The production of recombinant pharmaceutical proteins in plants. *Nature Reviews Genetics* 4: 794-805.

Nester, E.W., Gordon, M.D., Amasino, R.M., Yanofsky, M.F. 1984. Crown gall: a molecular and physiological analysis. *Annual Review of Plant Pathology* 35:387-413.

Peterson, R.K.D., Arntzen, C.J. 2004. On risk and plant-based biopharmaceuticals. *Trends in Biotechnology* 22:64-66.

Rogers, K.K. 2003. The Potential of Plant-Made Pharmaceuticals. http://www.plantpharma.org/ials/index.php?id=1

Sala, F., Rigano, M. M., Barbante, A., Basso, B., Walmsley, A.M., Castiglione, S. 2003. Vaccine antigen production in transgenic plants: strategies, gene constructs and perspectives. *Vaccine* 21:803-808.

Thomas, B.R. Van Deynze, A., Bradford, K.J. 2002. Production of Therapeutic Proteins in Plants. Agricultural Biotechnology in California Series 8078. University of California-Davis. 12pp.

Walmsley, A.M., Arntzen, C.J. 2000. Plants for delivery of edible vaccines. *Current Opinion in Biotechnology* 11:126-129.

Wolt, J.D., Peterson, R.K.D. 2000. Agricultural biotechnology and societal decision-making: the role of risk analysis. *AgBioForm* 3:39-46.

2

Statistical Data Mining-Concepts, Tools and Techniques for Bioinformatics

Ravi R. Saxena, Ritu R. Saxena, Akash Solanki and M.L. Lakhera

Today, when the collection of biological data has been increasing at explosive rate, the processing of these data is needed. It would be extremely valuable if the specialists who know what the data mean were also able to imagine ways to collect, process and exploit these data. Biological and agricultural scientists should note that concepts from computer science, discrete mathematics and statistics are being used increasingly to study and describe biological systems. The specialists, who know the subjects of mathematics, statistics and computer programming, are needed for solving the computational problems in biology. Statistical data mining approaches seem ideally suited for bioinformatics, since it is data rich, but lacks a comprehensive theory of life organization at molecular level, or in the words of I.E. Alcamo " Keeping up with the directions and application of DNA is a never ending job." Starting with possible definitions of statistical data mining and bioinformatics, this chapter will give a general side view of both the fields on the statistical data mining part and its technologies and discuss how statistical data mining may be used in bioinformatics.

STATISTICS AND ITS ESSENTIALITY

Statistics is the science of "learning from data". It includes everything from planning for the collection of data and subsequent data management to end-of-the-line activities, such as drawing inferences from numerical facts called data and presentation of results. Statistics is concerned with one of them basic of human needs: the need to find out more about the world and how it operates in face of variation and uncertainty. Because of the increasing use of statistics, it has become very important to under stand and practice statistical thinking or, in the words of H.G. Wells: "Statistical thinking will one day be as necessary for efficient citizenship as the ability to read and write".

DEFINING DATA MINING

Data mining is an analytic process designed to explore large amounts of data in search of consistent patterns and/or systematic relationships between variables, and then to validate the findings by applying the detected patterns to new subsets of data."

According to M. J. A. Berryand G. S. Linoff "Data mining is the process of exploration and analysis, by automatic or semiautomatic means, of large quantities of data inorder to discover meaningful patterns and rules."

"Data mining is the application of statistics in the form of exploratory data analysis and predictive models to reveal patterns and trends in verylarge data sets." ("Insightful Miner 3.0 User Guide")

"Data mining is finding interesting structure (patterns, statistical models, relationships) in databases."

DATA MINING IS USED

1. Detect patterns in events and behaviors.
2. Optimize product performance and manufacturing processes.
3. Data mining can be utilized in any organization that needs to find patterns or relationships in their data for example agriculture, horticulture, biotechnology etc wherever the derived insights will deliver scientific value.

In other words, data mining is the process of discovering valid, previously unknown, and ultimately comprehensible information from large stores of data. Data Mining is the process of extracting knowledge hidden in large volumes of raw data. Data mining automates the process of find in relationships and patterns in raw data and delivers results that can be either utilized in an automated decision support system or assessed by a human analyst. Modern computer data mining systems self-learn from the previous history of the investigated system, formulating and testing hypotheses about the rules which this system obeys. In general, data mining techniques can be divided into two broad categories: predictive data mining and discovery data mining (Mamcenko and Kulvietiene, 2004). *Predictive data mining* is applied to a range of techniques that find relationship between specific variable (called target variable) and the other variables in your data. The following are examples of predictive mining techniques: Classification is about assigning data records into pre-defined categories. In this case the target variable is the category and the techniques discover the relationship between the other variables and the category. Regression is about predicting the value of a continuous variable from the other variables in a data record. The most familiar value prediction techniques include linear and polynomial regression. *Discovery data mining* is applied to range of techniques that find patterns inside your data without any prior knowledge of what patterns exist. The following are examples of discovery mining techniques: Clustering is the term for range of techniques, which attempts to group data records on the basis of how similar they are. Association and sequence analysis describes a family of techniques that determines relationship between data records. The particularity of contemporary requirements for the data processing is the following: the data have the unlimited quantity, the data differ (quantitative, qualitative and textual), the results should be particular and comprehensible and the tools for the processing of raw data should be easy to use. The modern technology of Data Mining (discovery-driven data mining) is based on the concept of patterns, reflecting fragments of polydimensional relationships within the data. These patterns are regularities, which are characteristic to the data sub-retrievals that can be reflected compactly in the form, which is easy to comprehend for the human. The search for the patterns is carried out by means of techniques, which are not limited by prior proposals about the structure of retrieval and type of the value distribution of indicators to be analyzed (Shekhar and Chawla, 2003).

THE TYPICAL GOALS OF DATA MINING ARE

1. Identification of groups, clusters, strata, or dimensions in data that display no obvious structure.
2. Identification of factors that are related to a particular outcome of interest (root-cause analysis).
3. Accurate prediction of outcome variable (s) of interest.

MODELS FOR DATAMINING:

In the data mining literature, various "general frameworks" have been proposed to serve as blue prints for how to organize the process of gathering data, analyzing data, disseminating results, implementing results, and monitoring improvements.

1. CRISP mid-1990s by a European consortium of companies to serve as a non-proprietory standard process model for data mining.

Scientific Understanding ↔ Data Understanding

↓

Data Preparation → Modeling

↓

Evaluation

↓

Deployment

2. DMAIC Six Sigma methodology – data-driven methodology for eliminating defects, waste, or quality control problems of all kinds.

   Define → Measure → Analyze → Improve → Control

3. SEMMA (SAS Institute) –focused more on technical aspects of datamining. Sample → Explore → Modify → Model → Assess.

STEPS IN DATAMINING

a) Formulation of the problem

Before opening a software package and running an analysis, the analyst must be clear as to what question he wants to answer otherwise, it will be wastage of time and money.

b) Data preparation

This stage usually starts with data preparation that may involve the "cleaning" of the data (e.g., identification and removal of incorrectly coded data, etc.), data transformations, selecting subsets of records and, in the case of datasets with large numbers of variables ("fields") and performing preliminary feature selection. Data description and visualization arekey components of this stage (e.g. descriptive statistics, correlations, scatter plots, box plots, etc.).

i) **"Cleaning"of data**: Identification and removal of incorrectly coded data, Male=Yes, Pregnant=Yes

ii) **Datatransformations**: Data may be skewed (that is, outliers in one direction or another may be present). Log transformation, Box-Cox transformation etc.

iii) **Datareduction**: Selecting subsets of recordsand in the case of data sets with large numbers of variables ("fields"), performing preliminary feature selection.

iv) **Data description and visualization**: are key components of thisstage (e.g. descriptive statistics, correlations, scatter plots, box plots, brushing tools, etc.) Data description allows you toget a snapshot of the important characteristics of the data (e.g. central tendency and dispersion). Patterns are often easier to perceive visually than with lists and tables of numbers.

c) Model building and validation.

This stage involves considering various models and choosing the best one based on their predictive performance.

i) Data Mining involves creating models of reality.

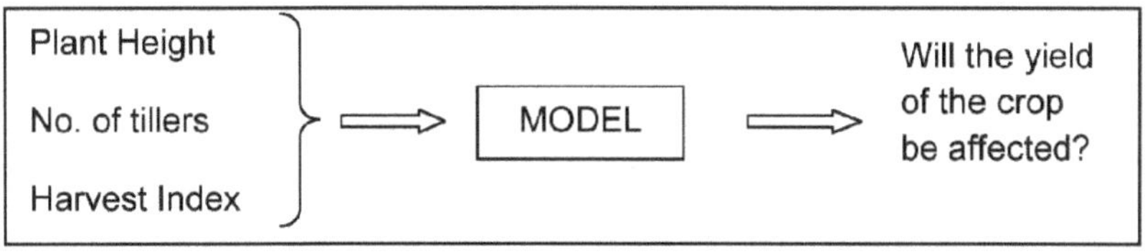

ii) A model takes one or more inputs and produces one or more outputs.

iii) A model can be"transparent", for example, a series of if/then statements where structure is easily discerned, or a model can be seen as a black box, for example, neural network, where the structure or the rules that govern the predictions are impossible to fully comprehend.

iv) A model is typically rated according to 2 aspects:

- Accuracy
- Understandability

v) These aspects sometimes conflict with one another.

vi) Decision trees and linear regression models are less complicated and simpler than models such as neural networks, boosted trees, etc. and thus easier to understand, however, you might be giving up some predictive accuracy.

vii) Remember not to confuse the data mining model with reality (a road map is not a perfect representation of the road) but it can be used as a useful guide.

viii) Validation of the model requires that you train the model on one set of data and evaluate on another independent set of data.

ix) There are two main methods of validation

- Split data into train/test data sets (75-25 split)
- If you do not have enough data to have a holdout sample, then use v-fold cross validation.

Ingeneral, the term over fitting refers to the condition where a predictive model (e.g., for predictive data mining) is so"specific" that it reproduces various idiosyncrasies (random"noise" variation) of the particular data from which the parameters of the model were estimated; as a result, such models often may not yield accurate predictions for new observations (e.g., during deployment of a predictive data mining project). Often, various techniques such as cross-validation and v-fold cross-validation are applied to avoid over fitting.

How predictive is the model?

Compute error sums of squares (regression) or confusion matrix (classification).

1. Error on training data is not a good indicator of performance on future data. The new data will probably not be exactly the same as the training data.
2. Overfitting: Fitting the training data too precisely. Usually leads to poor results on new data.

MODEL VALIDATION MEASURES

Possible validation measures are:-

i) Classification accuracy

ii) Total cost/benefit– when different errors involve different costs

iii) Lift and Gain scurves

iv) Error in Numeric predictions

Errorrate

i) Proportion of errors made over the whole set of instances.

ii) Training set error rate: is way too optimistic.

iii) We can find pattern seven in random data.

d) Deployment

When the goal of the data mining project is to predict or classify new cases (e.g., to predict the credit worthiness of individuals applying for loans), the third and final stage typically involves the application of the best model or models (determined in the previous stage) to generate predictions.

- A model is built once, but can be used over and over again.
- Model should be easily deployable.
  - A linear regression is easily deployed. Simply gather the regression coefficients.

- For example, if a new observed data vector comes in $\{x_1, x_2, x_3\}$, then simply plug into linear equation to generate predicted value,

Prediction $=b_0+ b_1x_1+ b_2x_2+ b_3x_3$

- A Classification and Regression Tree model is easily deployed: A series of If/Then/Else statements.

DATA MINING TASKS

Let us briefly define the main tasks well-suited for data mining, all of which involve extracting meaningful new information from the data. The six main activities of data mining are:

Classification: (examining the feature of a newly presented object and assigning it too neofaprede finedset of classes);

Estimation: (given some input data, coming up with a value for some unknown continuous variable);

Prediction: (the same as classification and estimation except that the records are classified according to some future behavior or estimated future value);

Affinity Grouping or Association: *rules* (determine which things go together, also known as dependency modelling);

Clustering: (segmenting a population into a number of subgroup sorclusters); and *description and visualisation* (exploratory or visual data mining).

Learning from data comes in two flavours: directed ("supervised") and undirected ("unsupervised") learning.

The first three tasks–classification, estimation and prediction–are all examples of supervised learning. In supervised learning ("class prediction") the goal is to use the available data to build a model that describes one particular variable of interest interm of the rest of the available data.The next three tasks–affinity group in association rules, clustering, and description and visualization, are examples of unsupervised learning. In unsupervised learning ("class discovery") no variable is singled out as the target; the goal is to establish some relationship among all the variables. Unsupervised learning attempts to find patterns or similarities among groups of records without the use of a particular target field or collection of pre-defined classes.

SUPERVISED LEARNING

Acategory of data mining methods that use a set of labeled training examples (e.g., each example consists of a set of values on predictors and outcomes) to fit a model that later can be used for deployment.

- Classification: response is categorical.

- Regression: response is continuous.
- Time Series: dealing with observations across time.
- Optimization: minimize or maximize some characteristic.

UNSUPERVISED LEARNING

A datamining method based on training data where the outcomes are not provided.

- Principal Component Analysis: feature reduction.
- Clustering: grouping like objects together.
- Association and Link Analysis: descriptive approach to exploring data that helps identify relationships among values in a database, e.g. Market basket analysis, those customers that buy hammers also buy nails. Examine conditional probabilities.

OVERVIEW OF DATA MINING TECHNIQUES

The commonly used data mining techniques below at a high level, focusing on thebig picture so that you can seehow each technique fits into the overall landscape.

- Descriptive Statistics
- Linear and Logistic Regression
- Analysis of Variance (ANOVA)
- Discriminant Analysis
- Decision Trees
- Clustering Techniques (KMeans & EM)
- Neural Networks
- Association and Link Analysis
- MSPC (Multivariate Statistical Process Control)

DESCRIPTIVE STATISTICS

Typically there is so much data in a database that it first must be summarized in order to make any sense of it. The first step in Data Mining is describing and summarizing the data. Two main types of statistics commonly used to characterize a distribution of data are:

i) Measures of Central Tendency:- Mean, Median, Mode

ii) Measures of Dispersion:- Standard Deviation, Variance

Visualization of the data is crucial. Patterns that can be seen by the eye leave a much stronger imprint than a table of numbers or statistics.

LINEAR AND LOGISTIC REGRESSION

Regression analysis is a statistical methodology that utilizes the relationship between two or more quantitative variables, so that one variable can be predicted from the others.

We shall first consider linear regression. This is the case where the response variable is continuous. In logistic regression the response is dichotomous.

Examples Include

i) Sales of a product can be predicted utilizing the relationship between sales and advertising expenditures.

ii) The performance of an employee on a job can be predicted by utilizing the relationship between performance and a battery of aptitude tests.

The size of vocabulary of a child can be predicted by utilizing the relationship between size of vocabulary and age of child and amount of education of the parents.

The simplest form of regression contains one predictor and one response.

$$Y = \beta_0 + \beta_1 X$$

The slope and intercept terms are found such that sum of squared deviations from the line are minimized. This is the principle of least squares.

What if the pattern in the data is NOT linear?

More predictors can be used.

i) Transformations.

ii) Interactions and higher order polynomial terms can beadded (hence linear regression does not mean linear in the predictors, but rather linear in the parameters), that is, we can easily bend the line into curves that are nonlinear.

iii) We can really develop complicated models using this approach; however, this takes expertise on the part of the modeler in both the domain of application as well as with the methodology. Techniques that we will look at later can do a lot of the "grunt" work for us.

What if the response is dichotomous?

We might use linear regression to predict the 0/1 response. However, there are problems such as:

If you use linear regression, the predicted values will become greater than one and less than zero if you move far enough on the X-axis. Such values are theoretically in admissible.

One of the assumptions of regression is that the variance of Y is constant across values of X (homoscedasticity).

There are a variety of regression applications in which the response variables only has two qualitative outcomes (male/female, default/no default, success/failure).

In a longitudinal study of coronary heart disease as a function of age, gender, smoking history, cholesterol level, and blood pressure, the response variable was defined to have to two possible outcomes: person developed heart disease; person did not develop heart disease during the study.

ANOVA

Analysis of Variance (ANOVA) allows you to test differences among means. The explanatory variables in an ANOVA are typically qualitative (gender, geographic location, plant shift, etc. If predictor variables are quantitative, then no assumption is made about the nature of the regression function between response and predictors. Some examples include:

i) An experiment to study the effects of five different varieties of a crop.

ii) An experiment to assess the effects of different amounts of a particular psychedelic drug on manual dexterity.

DISCRIMINANT ANALYSIS:

Discriminant function analysis is used to determine which variables discriminate between two or more naturally occurring groups. For example, an agricultural researcher may want to investigate which variables discriminate between high yielding varieties.

i) To tolerate water deficiency.

ii) To give high protein contents, or

iii) To disease resistance.

For that purpose the researcher could collect data on numerous variables prior to grow the crop. After germination, most plants will naturally fall into one of the three categories. Discriminant Analysis could then be used to determine which variable(s) are the best predictors.

A medical researcher may record different variables relating to patients' backgrounds in order to learn which variables best predict whether a patient is likely

i) To recover completely (group 1),

ii) Partially (group 2), or

iii) Not a tall (group3).

DECISION TREES:

Decision trees are predictive models that can be viewed as a tree. Models can be made to predict categorical or continuous responses. A decision tree is nothing more than a sequence or if/then statements.

CLUSTERING

Clustering is theme; a grouping of a number of similar things "a bunch of tree", "a cluster of admirers".

K Means Clustering: The classic k-Means algorithm was popularized and refined by Hartigan (1975; see also Hartigan and Wong, 1978). The basic operation of that algorithm is relatively simple: Given a fixed number of (desired or hypothesized) k clusters, assign observations to those clusters so that the means across clusters (for all variables) are as different from each other as possible.

EM Clustering: The EM algorithm for clustering is described in detail in Witten and Frank (2001). The goal of EM clustering is to estimate the means and standard deviations for each cluster, so as to maximize the likelihood of the observed data (distribution). Put another way, the EM algorithm attempts to approximate the observed distributions of values based on mixtures of different distributions in different clusters.

NEURAL NETWORKS

Like most statistical models neural networks are capable of performing three major tasks including *regression*, *classification*. Regression tasks are concerned with relating a number of input variables x with set of continuous *outcomes* (targetvariables). By contrast, classification tasks assign class memberships to a categorical target variable given a set of input values. In the next section we will consider regression in more details.

MSPC

i) Built upon the capabilities of *PCA* and *PLS* techniques, *MSPC* is a selection of methods particularly designed for process monitoring and quality control in industrial batch processing.

ii) Batch processes are of considerable importance in making products with the desiredspecificationsandstandardsinmany sectors of theindustry, such as herbicides, insecticides etc.

iii) The objectives of batch processing are related to profitability achieved by reducing product variability as well as increasing quality.

TOOLS FOR DATA MINING IN BIOINFORMATICS

Molecular biology has witnessed an information revolution as a result both of the development of rapid DNA sequencing techniques and of the corresponding progress in computer –based technologies, which are allowing us to cope with this information deluge increasing efficient ways. The broad term that was coined in the mid- 1980s to encompass computer application inbiological sciences is bioinformatics. This term bioinformatics has been commandeered by several different disciplines to mean rather different things. In its broadest sense the term can be considered to mean information technology applied to the management and analyses of biological data; this hyas implication in diverse areas ranging from artificial intelligence and robotics to genome analysis. In the context of genome initiatives, the term was originally applied to computational manipulation and analysis of biological sequences data (DNA and /or Protein).

More generally:

i) "Bioinformatics is the science of managing and analysing biological data using advanced computing techniques. Especially important in analysing genomic research data" (Glossary of genetic terms from the "DOE Human Genome Program")

ii) "Bioinformatics is about searching biological databases, comparing sequences, looking at protein structures, and more generally, asking biological questions with acomputer." (J.M. Claverieand C. Notredame)

iii) "Bioinformatics is the study of genetic and other biological information using computer and statistical techniques." (L. Helmuth)

SAGE MAP

Provides a tool for performing statistical tests designed specifically for differential-type analyses of SAGE (Serial Analysis of Gene Expression) data. The data include SAGE libraries generated by individual labs as well as those generated by the Cancer Genome Anatomy Project (CGAP), which have been submitted to Gene Expression Omnibus (GEO). Gene expression profiles that compare the expression in different SAGE libraries are also available on the Entrez GEO Profiles pages. It is possible to enter a query sequence in the SAGEmap resource to determine what SAGEtags are in the sequence, then map to associated SAGEtag records and view the expression of those tags in different CGAP SAGE libraries.

BASIC LOCAL ALIGNMENT SEARCH TOOL (BLAST)

For comparing gene and protein sequences against others in public databases now comes in several types including PSI-BLAST, PHI-BLAST, and BLAST 2 sequences. Specialized BLASTs are also available for human, microbial, malaria, and other genomes, as well as for vector contamination, immunoglobulin and tentative human consensus sequences.

ENTREZ GENE

Each Entrez Gene record encapsulates a wide range of information for a given gene and organism. When possible, the information includes results of analyses that have been done on the sequence data. The amount and type of information presented depend on what is available for a particular gene and organism and can include: (1) graphic summary of the genomic context, intron/exon structure, and flanking genes, (2) link to a graphic view of the mRNA sequence, which in turn shows biological features such as CDS, SNPs, *etc*., (3) links to gene ontology and phenotypic information, (4) links to corresponding protein sequence data and conserved domains, (5) links to related resources, such as mutation databases. Entrez Gene is a successor to Locus Link.

MODEL MAKER

Allows you to view the evidence (mRNAs, ESTs, and gene predictions) that was aligned to assembled genomic sequence to build a gene model and to edit the model by selecting or removing putative exons. You can then view the mRNA sequence and potential ORFs for the edited model and save the mRNA sequence data for use in other programs. Model Maker is accessible from sequence maps that were analyzed at NCBI and displayed in Map Viewer.

BLINK ("BLAST LINK")

Displays the results of BLAST searches that have been done for every protein sequence in the Entrez Proteins data domain.

TAXPLOT

A tool for 3-way comparisons of genomes on the basis of the protein sequences they encode. To use TaxPlot, one selects a reference genome to which two other genomes are compared. Pre-computed BLAST results are then used to plot a point for each predicted protein in the reference genome, based on the best alignment with proteins in each of the two genomes being compared.

COGS (CLUSTERS OF ORTHOLOGOUS GROUPS)

A natural system of gene families from complete genomes. Clusters of Orthologous Groups (COGs) were delineated by comparing protein sequences encoded in 43 complete genomes, representing 30 major phylogenetic lineages. Each COG consists of individual proteins or groups of paralogs from at least 3 lineages and thus corresponds to an ancient conserved domain.

MAP VIEWER

Shows integrated views of chromosome maps for many organisms, including human and numerous other vertebrates, invertebrates, fungi, protozoa, and plants. Map Viewer is used to view assembled genomes (either draft or complete) and is a valuable tool for the identification and localization of genes and other biological features. Multiple map displays are aligned based on shared marker and gene names when available, and sequence map displays are based on a common sequence coordinate system. Sequence data for chromosome regions of interest can be downloaded, biological annotations can be viewed in graphical format and/or downloaded in tabular format, and gene models can be manipulated in the associated Model Maker tool.

REFERENCES

Baldi, P. and Brunak, S. 2001. Bioinformatics-The Machine Learning Approach (2ndEd.). Cambridge, MA: MIT Press.

Berry, M.J.A. and Linoff, G.S. 1997. Data Mining Techniques for Marketing, Salesand Customer Support. New York: Wiley.

Claverie, J.M. and Notredame, C. 2003. Bioinformatics for Dummies. NewYork:Wiley.

Demidov, V.V. 2003. Golden Jubilee of the DNA double helix. Trends in Biotechnology, 21:139-140.

Dutton, G. 2002. Geneexpression data mining. *The Scientist*, 16:50.

Ewens, W. and Grant, G. R. 2001. Statistical MethodsinBioinformatics: An Introduction. New York: Springer.

Han, J. 2002. Howcandatamininghelpbio-dataanalysis? *In:* M.J. Zaki, J.T.L. Wangand H.T.T. Toivonen (Eds.), Proceedings of the "2nd ACM SIGKDD Workshop on Data Mining in Bioinformatics", 1-2 (www.cs.rpi.edu/~zaki/BIOKDD02/).

Han, J. and Kamber, M. 2000. Data Mining: Conceptsand Techniques. Morgan Kaufmann Publishers. Hand, D. J., Mannila, H. and Smyth, P. 2001. Principles of Data Mining. Cambridge, MA: MIT Press.

Hastie, T., Tibshirani, R. and Friedman, J.H. 2001. The Elements of Statistical Learning: Data Mining, Inference, and Prediction. New York: Springer.

Holmes, S.P. 1999. Phylogenies: anoverview. In: M.E. Halloran and S. eisser (Eds.), Statistics in Genetics, The IMA Volumes in Mathematics and its Applications, 112:81-118.

Lange, K. 2002. Mathematical and Statistical Methods forGenetic Analysis (2ndEd.).New York: Springer. Liu, J. S. 2002. Bioinformatics: microarrays and beyond. Amstat News, 303:59-67.

Mann, B., Williams, R., Atkinson, M., Brodlie, K, Storkey, A. and Williams, C. 2003. Scientific data mining, integration, and visualisation. Report of the workshop held at the e-Science Institute, Edinburgh, 24-25 October2002. (umbriel.dcs.gla.ac.uk/NeSC/general/talks/sdmiv/report.pdf).

Nguyen, D.V., Arpat, A.B., Wang, N. and Carroll, R.J. 2002. DNA microarrayexperiments: biological and technological aspects.Biometrics, 58:701-717.

Sebastiani, P., Gussoni, E., Kohane, I.S. and Ramoni, M.F. 2003. Statistical challenges in functional genomics (with discussion). Statistical Science, 18:33-70.

Waterman, M. S. (1995). Introduction to Computational Biology: Maps, Sequences and Genomes. London: Chapman & Hall.

Witten, I. H. and Franke, E. 2000. Data Mining: Practical Machine Learning Tools and Techniques with Java Implementations. Morgan Kaufmann Publishers.

"Bioinformatics World" magazine: www. bioinformaticsworld. info

"Functional Genomics: Beyond the Genome– Introduction to Bioinformatics and Genomics": www.library.csi.cuny.edu/~davis/Bioinformatics/Bioinfo_Genomics. html

Homepage of H. Jiawei (data miningand bioinformatics): www-faculty.cs.uiuc.edu/~hanj

Homepage of A. Robinson (visualisation, dataminingand bioinformatics): industry.ebi.ac.uk/~alan

Homepage of M.J. Zaki (dataminingand bioinformatics): www.cs.rpi.edu/~zaki

http://www.ncbi.nlm.nih.gov/tools.

3

Prediction of 3D – Structure of Proteins Using Artificial Neural Networks

K. Meena

The development of bio-informatics started with the working of computers and accumulation of data on genes and proteins. There are 100,000 proteins in the human body or 1 million. In the last decade the number of protein sequences gathered in databases increased tremendously, but the key to understand the function of a protein is not its sequence but its structure. Experimental determination of 3 dimensional native folded structure of a protein usually can be done via crystallography techniques. But such experiments are time consuming. Currently there are hundreds of different fully automated approaches to protein 3D structure prediction.

All proteins are polymers of Amino acids. The polymers also known as polypeptides consist of a sequence of 20 different L-μ-amino acids also referred to as residues. For chains under 40 residues the term peptide is frequently used instead of protein. To be able to perform their biological function, proteins fold into one that is driven by a number of non covalent interactions such as hydrogen bonding, ionic interaction, Vander Waal's forces and hydrophobic packing. In order to understand the functions of proteins at a molecular level, it is often necessary to determine the three dimensional structure of proteins. The modern Information Technology techniques like Artificial Intelligence have opened new pathways to researchers. Once ANN has been trained it is capable of producing accurate output for the unknown inputs. Many researchers have tried to analyze the large sequence sized gene to determine the protein, predict the secondary structure of the protein and model its three dimensional (3D) view [11].

In this study, two different ANN architectures namely Sigmoidal networks and Radial Basis Function (RBF) networks are used to predict residue contacts of a protein in order to obtain structural constraints for the three-dimensional structure. These two ANN architectures are compared based on their performance. ANN is more suitable for the protein structure prediction and it provides the accuracy level of 80%. This study aims at predicting and understanding the protein structure from their contact maps.

Amino acids are the building blocks of proteins. An amino acid is any molecule

Structure and Properties of Amino ACIDS

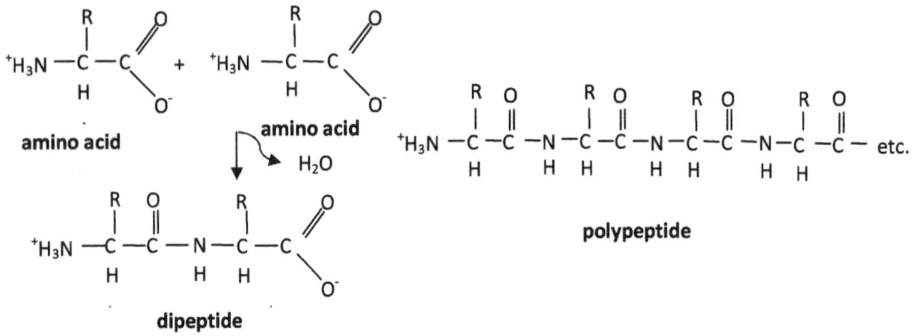

Fig. 1: Protein structure

that contains both amine and carboxylic acid functional groups as represented in Figure (1). Amino acids are the basic structural building units of proteins and form short polymer chains called peptides or polypeptides, which in turn form structures called proteins and is illustrated in Figure 3. The process of such formation from an mRNA template is known as translation, which is part of protein synthesis.

STANDARD AMINO ACID PHENYLALANINE

Phenylalanine is one of the standard amino acids. Twenty amino acids are encoded by the standard genetic code and are called *proteinogenic* or *standard amino acids*. Combinations of these amino acids produce every single essential protein for the homeostasis of the human body. In addition to protein synthesis, amino acids have other biologically-important roles. Glycine and glutamate are neurotransmitters as well as standard amino acids in proteins.

Some of the 20 standard amino acids are called essential amino acids because they cannot be synthesized by the body from other compounds through chemical reactions, but instead must be taken in with food. In humans, the essential amino acids are lysine, leucine, isoleucine, methionine, phenylalanine, threonine, tryptophan, valine. Histidine and arginine are generally considered essential only in children, because of their inability to synthesise them. The phrase "branched-chain amino acids" is sometimes used to refer to the aliphatic amino acids: leucine, isoleucine and valine.

OCCURRENCE AND FUNCTIONS

Proteins are found in all living cells. They form essential constituents of protoplasm, cell membrane, or complex protein combined with lipids and nucleic acids. Genes, which carry hereditary characteristics, are nucleic proteins. Where they occur, they play vital role in the biological functions, and serve as buildings blocks for cellular and organic structures. Proteins from components of catalysts are called enzymes, which are essential for the metabolic reactions.

They are also constituents of most of the hormones, which are regulators of the metabolism of important physiological compounds. Proteins form about 15% of the total body weight of the average adult. The proteins in different tissues such as connective tissue, bones, muscles, brain, blood and other biological fluids differ in composition and properties. The protein content of plants is relatively less than that of animal. In the plants, seeds and cereals contain the highest protein content, while the leaves, stem and tubers contain much less protein.

The physiological significance and functions of proteins are varied but vital. They regulate and integrate the numerous physiological and metabolic processes in the body through hormones, enzymes and nucleoproteins. The plasma proteins help in the maintenance of acid-base balance and osmotic pressure of the body fluids.

The general structure of proteinogenic alpha amino acids is where R represents a side chain specific to each amino acid. Amino acids are usually classified by the properties of the side chain into four groups. The side chain can make them behave like a weak acid, a weak base, a hydrophile, if they are polar, and hydrophobe if they are nonpolar (See Figure 2).

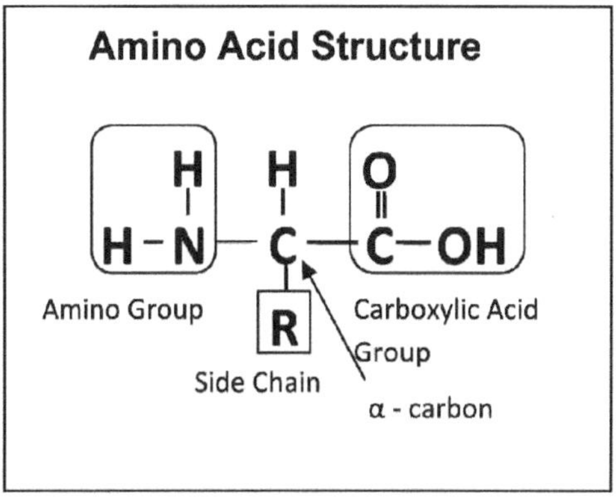

Fig. 2. General Structure of Amino Acids

FOUR LEVELS OF ORGANIZATION

Protein structure can be classified into four levels of organization. They are described as primary structure, secondary structure, tertiary structure and quarternary structure. Different visualizations in Figure 2 show various aspects of structure.

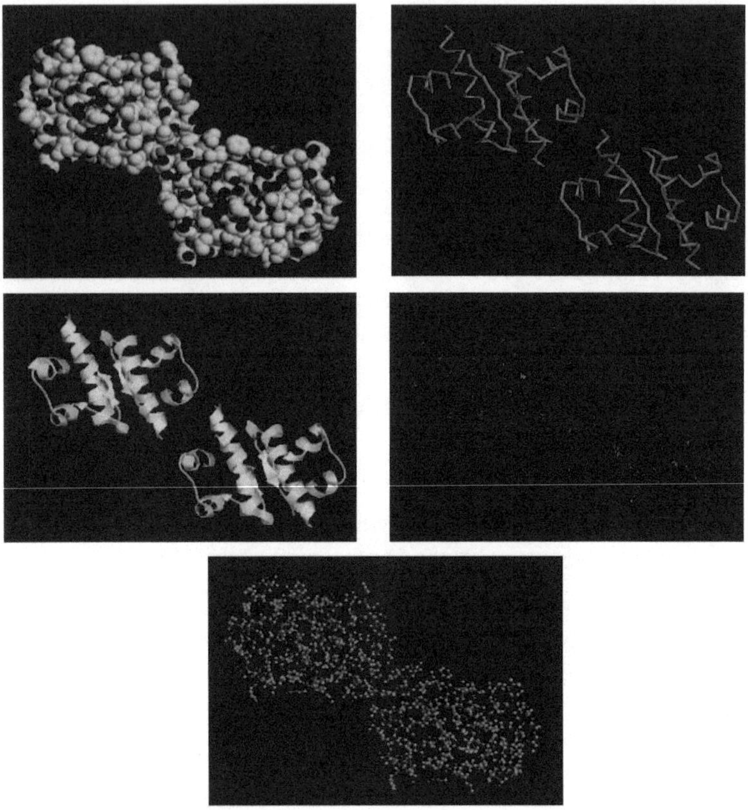

Fig. 3: Protein structure representations

PRIMARY STRUCTURE

The primary structure of a protein molecule refers to the sequence of amino acids in the polypeptide chain. The amino acids are linked by covalent peptide bonds. The amino acid sequence which is coded by DNA determines the final three dimensional form adopted by the protein in its native state. It includes the location of the disulphide (cystine) bonds.

SECONDARY STRUCTURE

The next level of protein structure generally refers to the amount of structural regularity or shapes that the polypeptide chain adopts. The proteins generally have a mix of different kinds of secondary structure. The peptide chain formed by the primary structure assumes a three dimensional secondary structure by folding or coiling, consisting of specific helically coiled structure in a regular manner. The coiling is the result of linking of the carboxyl group and the amino group of the neighbouring amino acid residue of the peptide chain, by hydrogen bonds and disulphide bonds.

The hydrogen bonds are formed when a hydrogen atom (H) is shared between the oxygen (O) of the CO and nitrogen (N) of NH of different peptide bonds. The secondary structure is formed by the ordering of the polypeptide chain as dictated by the DNA encoded primary sequence. Secondary structure chains tend to be located on the surface of the molecule in contrast with water molecule.

TERTIARY STRUCTURE

Three-dimensional structure of a single protein molecule; a spatial arrangement of the secondary structures. The Tertiary structure is a global ordering in all angles and atoms.

QUATERNARY STRUCTURE

When a protein consists of two or more peptide chains held together by non-covalent interactions or by covalent cross link, is referred to as Quartenary Structure. In Quartenary structure, there are several monomeric units and the primary, secondary, tertiary structures may combine. The polymerization of these monomers or subunits is described as quartenary structure. Glutamic dehydrogenase is a tetramer and aldolase molecule is trimer.

TOOLS FOR PROTEIN SEQUENCE ANALYSIS

Sequence analysis tools include:DNA -> Protein [Translate]

- Pattern and profile searches.
- Post-translational modification and topological prediction.
- Primary structure analysis.
- Structure prediction (2D and 3D).
- Alignment.

Different protein analysis systems are as follows:

Expasy server: (Expert Protein Analysis System) is dedicated to the analysis of protein sequences and structures: http://au.expasy.org

PredictProtein:A service for sequence analysis, and structure prediction: http://www.predictprotein.org/newwebsite/submit.html

TMpred:http://www.ch.embnet.org/software/TMPRED_form.html

TMHMM: Predicts transmembrane helices in proteins (CBS; Denmark): http:// www.cbs. dtu.dk/ services/TMHMM-2.0/

big-PI: Predicts GPI-anchor site: http://mendel.imp.univie.ac.at/ sat/gpi/ gpi_server.html

DGPI: Predicts GPI-anchor site: http://129.194.185.165/dgpi/ index_en.html

SignalP: Predicts signal peptide: http://www.cbs.dtu.dk/services/SignalP/

PSORT: Predicts sub-cellular localization:http://www.psort.org/

TargetP: Predicts sub-cellular localization: http://www.cbs.dtu.dk/services/ TargetP/

NetNGlyc: Predicts N-glycosylation sites: http://www.cbs.dtu.dk/services/ NetNGlyc/

PTS1: Predicts peroxisomal targeting sequences: http://mendel. imp.univie. ac.at/ mendeljsp/ sat/pts1/PTS1predictor.jsp

MITOPROT: Predicts of mitochondrial targeting sequences: http://ihg.gsf.de/ ihg/mitoprot.html

Hydrophobicity: http://www.vivo.colostate.edu/molkit/hydropathy/index.html

ARTIFICIAL NEURAL NETWORK APPROACH

Artificial Neural Networks (ANNs) have been applied to many aspects of predicting protein structure. At the beginning such methods were designed as a quick and dirty demonstration that Artificial Intelligence based approaches could solve real-life problems. However, more through investigations enabled researchers to include information used by experts into neural network-based tools. Several neural network-based methods have eventually contributed significantly to advancing the field of bioinformatics. Neural networks were first applied to the prediction of secondary structures [2, 6]. This stimulated many subsequent studies of neural networks [3, 4] in the Teritary structure prediction.

METHODS AND PROBLEM DEFINITION

Given the protein sequence, we want to find out its 3D structure. There are many types of ANN architectures and each of them has applications in different types of problems such as classification, pattern recognition and forecasting. The ANN with Sigmoidal and RBF networks are used in this work for finding 3D structure of proteins. Figure 4 gives the architectural view of the neural network architecture used in the system.This neural network contains three layers: input layer, hidden layer and output layer.

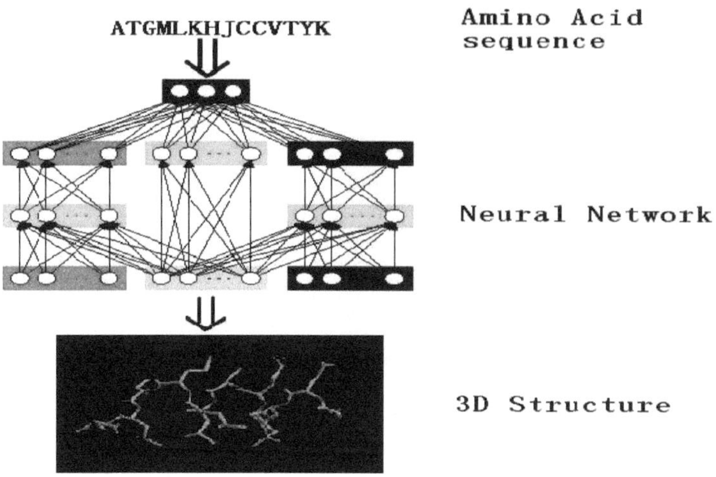

Fig. 4: Architectural view of ANN – Training Model

Input Layer: The input layer gets the input values to the neural network. There are twenty amino acids in a human body, each of these are assigned 5 bit encoding concept, amino acid pattern system assign the particular value and weights which can be any number. This layer contains five neurons or nodes.

Hidden Layer: The hidden layer or the system consists with a middle layer that contains twenty neurons. According to the knowledge and the results obtained from the past experience ten or more than ten neurons for the hidden layer are the optimal way to design this architecture.

Output Layer: Output layer contains five neurons or nodes. Output is in form of normalized contact map groups.

PROTEIN CONTACT MAP

A contact map is a particularly useful two dimensional representation of a protein's tertiary structure. A protein contact map represents the distance between every two amino acids of a three-dimensional protein structure in a two-dimensional matrix. It can be used to describe the similarity between protein structures and prediction of tertiary structure. These maps are either predicted from protein sequence or calculated from a given structure.

In the contact map, the primary sequence of the protein is placed in both rows and columns of the matrix. If two residues are near to each other within a specific cutoff radius, then, the entry in the contact map corresponding to these two residues is 1, otherwise it is 0. All short and long range interactions in a protein of known structure can be represented in its contact map. An example is shown in Figure 5.

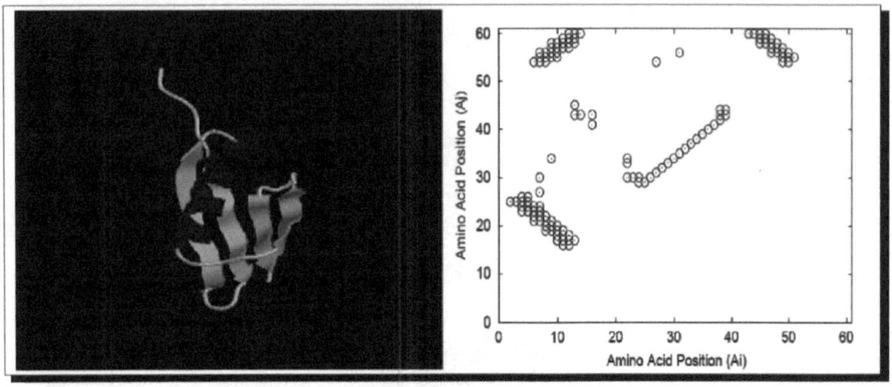

Fig. 5: Left: 3D structure for protein G. Right: Contact Map for protein G Circles indicate residue contacts, while non-contacts are represented by empty space.

The native structure of a protein is approximated by the set of the coordinates listed in the PDB (Protein Data Bank) file.

http://www.rcsb.org/pdb/
ftp://ftp.rcsb.org/pub/pdb/data/biounit/coordinates/all/
http://pir.georgetown.edu/pirwww/
http://www.sanger.ac.uk/Software/Pfam/
http://www.hprd.org/
http://pkr.sdsc.edu/html/index.shtml
http://www.hprd.org/
Fig.Flow chart of protein sequence analysis

In Figure 6, the different stages of protein sequence analysis are illustrated. ANNs accept the input as amino acid sequence and produces output in the form of contact map groups. The output contact map is constructed by considering the distance between amino acids in the protein structure.

FLOW CHART OF IMPLEMENTATION

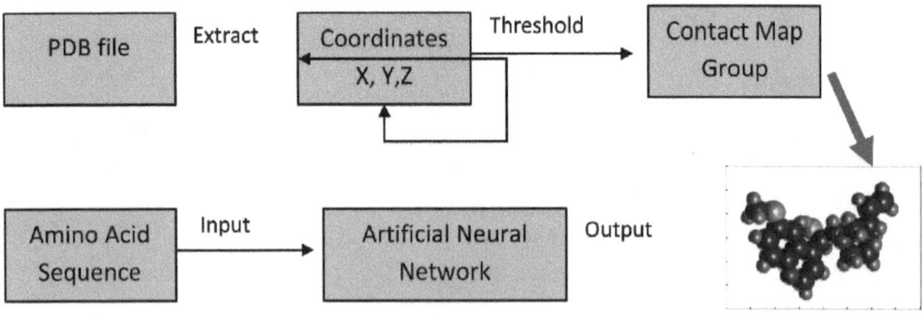

Fig. 6: Flow chart of protein sequence analysis

PREDICTION WITH ANNs

In this study, physical and chemical features of amino acids as well as other features involving the protein length and the primary sequence are used for predicting the contacting residues.

INPUT AND OUTPUT OF THE ANN

The input of the ANN is two residues or a window of residues, the length of the protein and the sequence separation of the corresponding residues (number of residues between them along the chain). The output of the network is the contact propensity of the corresponding residues. In other words, features of two residues are applied to the network and the desired action from the network is a prediction of these residues is in contact or not.

DATASETS

A dataset composed of 600 proteins used for this analysis is obtained from freely available online PDB (Protein Data Bank). The Protein Data Bank is the single worldwide repository for the processing and distribution of 3D biological macromolecular structure data. For selection of proteins, certain facts and constraints are considered for the better results. They are

i) Keeping global information of protein sequences to get better prediction.

ii) Proteins whose length less than or equal to 50 are chosen for easy testing.

The selected protein sequence data set is divided into three subsets for training, validation and test separately. The chosen data for input to ANNs are preprocessed using the following encoding scheme:

i) There are 20 amino acids, so we use 5 bit to encode each amino acid. Such as 00000 represent A, 00001 represent C,

ii) Use fixed length, say 60. And use 11111 to fill in the gaps.

TRAINING ANN

There is a training phase in ANNs at which the network receives a number of training patterns and adjusts its weights in order to attain corresponding outputs for each of the patterns. ANN trains protein sequence in the form of amino acids patterns. The ANN consists of one input layer, hidden layer and an output layer as shown in Figure 7. The input consists of amino acids and they can be represented as $x_1, x_2, \ldots, x_n$ which constitutes a protein sequence.

In this study, Backpropagation algorithm with the Sigmoidal Function and Radial Basis Function are used for training the ANN. These learning algorithms adjust the weights and biases of the network to minimize the mean square error, where the error is the difference between the target output and the network output. The mean square error calculated after each iteration, is called training error, and it tends to decrease throughout iterations. At this phase, the network learns rules in the training set and stores them in its weights and biases. After this training process, the network is ready to test whether it can produce reasonable outputs for the patterns not encountered in the training phase, which is called generalization.

Two different ANN architectures are used to predict contacts within proteins. The networks take a pair of residues whose contact propensity is under examination as an input and the output is the contact propensity of this pair which is a number between 0.1 and 0.9. The Sigmoidal and Radial Basis transfer functions are given by;

$$\text{Sigmoidal Function} = \frac{1}{1+e^{-n}}$$

$$\text{Radial Basis Function} = \frac{e^{-(x-c)^2}}{2\sigma^2}$$

Wheres is a standard deviation and c is an N – dimensional vector representing the center of RBF. By using these functions the ANNs are trained. The complete structures of these two ANNs are illustrated in Figure 7.

The output is received after the generation of weights using random number generation.

ILLUSTRATION

From a protein data base (PDB), protein sequences are taken as inputs in the form of coordinates x, y and z. The Contact map is computed using the equation;

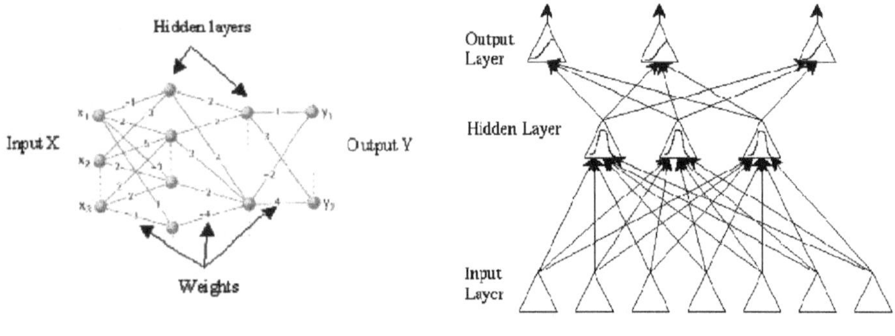

(a) Sigmoidal Network (b) RBF Network

Fig.7: Two different Artificial Neural Network architectures to train

$$D(i,j) = \sqrt{(x_i - x_j)^2 + (y_i - y_j)^2 + (z_i - z_j)^2}$$

Then, the obtained distance matrix is further normalized using the following equation;

$$D'(i,j) = \frac{D(i,j) - \min}{\max - \min} \quad D'(i,j) = \frac{D(i,j) - \min}{\max - \min}$$

Sample Contact map:

(j along the columns, i along the rows)

0.000	3.426	6.664	8.712
3.426	0.000	3.621	5.451
6.664	3.621	0.000	3.750
8.712	5.451	3.750	0.000

Normalized Contact map:

Normalized distance between [0, 1]

0.000	0.000	0.613	1.000
0.000	0.000	0.037	0.383
0.613	0.037	0.000	0.061
1.000	0.383	0.061	0.000

The result is a symmetric square N*N matrix whose elements are the distances among the amino acids in the matrix.

RESULTS AND PERFORMANCE COMPARISON

The earlier investigations using statistical and other techniques have achieved only a maximum accuracy level of 50%. ANNs, as prediction tools are used for several reasons. They are ANNs have a very good performance on prediction problems;are easy to implement and keep global information; are quite successful in problems where rules crucial to the required decision are subtle or deeply hidden.

The methodology used to evaluate the network performance is based on the number of correctly predicted contacting residues and number of false positives. Correct contacts (CC) is the ratio of number of actual contacting residue pairs whose network outputs are higher than the selected *threshold* to the total number of contacts. False positive (FP) is the ratio of number of actual non-contacting residue pairs whose outputs are higher than the selected *threshold* to total number of non-contacts. The aim of this study is to increase the number of correctly predicted contacts while decreasing the number of false positives as much as possible. The network outputs are the real numbers in the interval [0, 1] and higher the output, more probable that the input residue pair is contacting. In order to determine the correctly predicted contacts, a *threshold* is selected. The residue pairs whose outputs are equal to or higher than selected threshold are assigned as contacts and other pairs are assigned as non-contacts and the accuracy is determined.

ACCURACY (A) of the network is defined as the ratio of the correctly predicted contacts by the network to the actual number of contacts in a protein and calculated according to;

$$\text{Accuracy (A)} = \frac{N_c^*}{N_c}$$

Where N_c^* the number of is correctly predicted contacting residues by the network and N_c is the actual number of contacts within the protein.

It is observed form the results of experiments that the RBF networks with ten hidden nodes perform better than the sigmoidal networks. The performance comparison between Sigmoidal and RBF networks are illustrated in Table 1 and depicted in Figures 8(a), 8(b) and 8(c) respectively the comparison based on validation set 1, validation set 2 and all proteins.

Table 1: Performance comparison between Sigmoidal and RBF network

	Sigmoidal network with 10 hidden nodes			RBF network with 10 hidden nodes		
	CC (%)	FP (%)	<A>	CC (%)	FP (%)	<A>
Val Set 1	10.9	1.7	0.156±0.001	12.6	3.6	0.160±0.003
Val Set 2	4.5	0.7	0.085±0.001	7.3	1.3	0.094±0.002
All proteins	8.1	1.0	0.078±0.001	10.8	1.7	0.086±0.031

CC-Correct Contacts, FP – False Positives, A – Accuracy

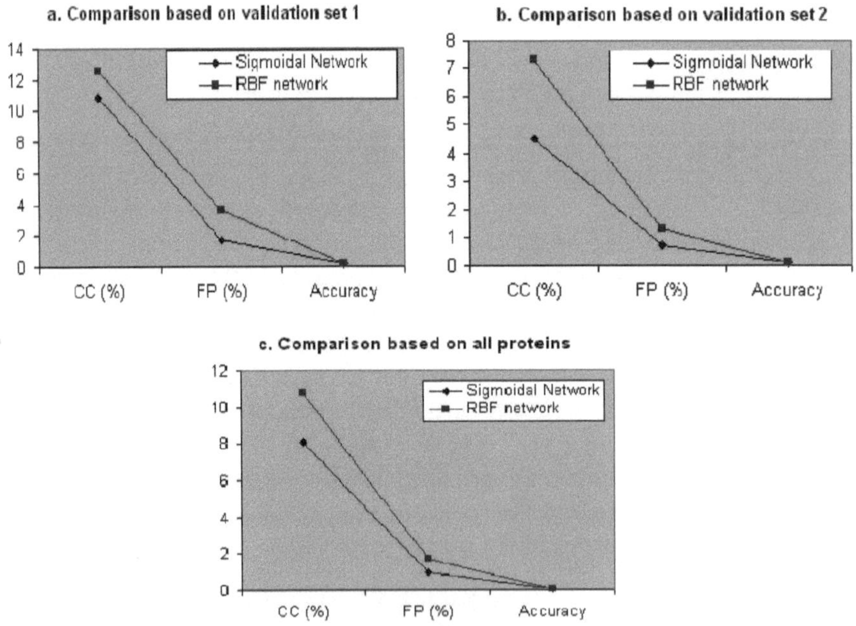

Fig. 8: Performance comparison between Sigmoidal and RBF networks

CONCLUSION

There are number of algorithms available to predict the protein structure. The accuracy of the predicted result is very low in all these techniques. Existing prediction processes achieve only 50% of accuracy. The concept of Artificial Neural Networks is used in the prediction system to maximize the accuracy rate. The prediction accuracy rates are approximately 75% for sigmoidal networks and 80% for RBF networks.

For the effective use of ANN, the compression of input data is needed to learn. It is recommended to compress input and output data small enough for an ANN to learn. A lengthy protein sequence of various lengths can be compressed to a protein structure of same length. For example, if ABCD is a small circle, then AAABBBCCCDDD is a bigger one. It is found that the Use of ANN in the prediction method increases the accuracy rate when compared to other methods. The accuracy rate of the predicted structure of proteins may also be increased further by pruning methodology or by adopting hybrid architecture.

REFERENCES

Arbib M. 1995. The handbook of brain theory and neural networks. Bradford Books/The MIT Press, Cambridge, MA.

Bohr, H., Bohr, J., Brunak, S,. Cotterill, R.M., Lautrup, B., Norskov, L., Olsen, O.H., Petersen S.B. 1988. Protein secondary structure and homology by neural networks. The alpha-helices in rhodopsin. *FEBS Lett.,* 241: 223-228.

Holley, H. and Karplus, M. 1989. Protein Secondary Structure Prediction with a Neural Network.*Proc. Natl Acad. Sci. USA,* 86:152-156.

Kneller, D., Cohen, F. and Langridge, R. 1990.Improvements in protein secondary structure prediction by an enhanced neural network. *Journal of Molecular Biology,* 214:171-182.

Kohonen T. 1982. Self-organized formation of topologically correct feature maps. *Biol. Cybern.,* 43:59-69.

Qian, N. and Sejnowski T. 1988.Predicting the secondary structure of globular proteins using neural network models. *Journal of Molecular Biology,* 202:865-884.

Rost, B. and Sander, C. 1993. Improved Prediction of Protein Secondary Structure by Use of Sequence Pro¯les and Neural Networks.*Proc. Natl Acad. Sci. USA,* 90:7558-7562.

Rost, B. and Sander, C. 1993. Prediction of protein secondary structure at better than 70% accuracy.*Journal of Molecular Biology,* 232:584-599.

Rost, B., Sander, C. and Schneider, R. 1994. PHD - an automatic server for protein secondary structure prediction. *CABIOS,* 10:53-60.

Rumelhart, D.E., McClell, J.L. and *Parallel distributed processing*. 1986. Explorations in the microstructure of cognition. *MIT Press,* Cambridge, MA., U.S.A.

Mount, D.W. 2004. Sequence and Genome Analysis, *Cold spring Harbor Laboratory Pres.*http://predictioncenter.org

Petersen, T.N., Lundegaard, C., Nielsen, M., Bohr, H., Brunak, S., Gippert, G.P. and Lund, O. 2000. Prediction of Protein Secondary Structure at 80% Accuracy. *Proteins*, 41:17-20.

4

A Genomic Approach to Study the Insect Mitochondrial Evolution by Non coding LSU and SSU rRNA

S. Gifta Lorraine, Christy Jancy Joy P., Arul Samraj D. and Priyakumari Joyce C.

Mitochondria are key energy generators in most eukaryotic cells. In recent years a number of mitochondrial genomes have been completely sequenced, contributing to the knowledge of molecular features related to function and evolution of this peculiar genome (Boore, 1999). The pathways of mitochondrial DNA evolution are obscured by their extraordinary size. Gene content, gene rearrangement and mode of expression are the characteristics of mitochondrial genomes. The mitochondrial genomes from eukaryotes reveal large size differences between animal, fungal and plant species. The average size of animal mtDNA is ~ 16 kb (Boore, 1999), whereas plant mtDNAs range from 200 to 2500 kb (Palmer, 1990). The size of fungal mtDNA's range from 19 kb to 176 kb (Hudspeth, 1995).The mtDNA genome has a limited number of protein coding genes such as NADH ubiquinone oxidoreductase (ND2 to ND5), cytochrome c oxidase (COX), succinate dehydrogenease (SDH), ATP synthase (ATP6), cytochrome (CYTB), etc. Among non-coding genes only rRNA and tRNA genes are found.Ribosomal RNA (rRNA) is a component of the ribosomes, the protein synthetic factories in the cell. rRNA molecules are extremely abundant. They make up at least 80% of the RNA molecules found in a typical eukaryotic cell. Synthesis of the three nucleolar rRNA molecules is unusual as they are made on one primary transcript that is chopped up into three mature rRNA molecules. These rRNA molecules and the 5S rRNA combine with the ribosomal proteins in the nucleolus to form pre 40 s and pre 60S ribosomal subunits. These pre-subunits are exported to the nucleus where they mature and assume their role in protein synthesis. The rRNA molecules have several roles in protein synthesis. The 28S rRNA having a catalytic role forms part of the peptidyl transferrase activity of the 60S subunit. The 18S rRNA is involved in correct positioning of the mRNA and the peptidyl tRNA.

Finally, the rRNA molecules having a structural role. They fold into three-dimensional shapes that form the scaffold on which the ribosomal proteins assemble. The aim of this study is to analyze rRNA non-coding sequences of the mtDNA of eleven species of insect and to determine the role of conserved sequences across the eight insect orders to show the mtDNA evolutionary origin. RNA functions as an information carrier, catalyst and regulatory element, perhaps reflecting its importance in the earliest stages of evolution. The structures of RNAs provide insight into the mechanisms behind these functions. The ribosomal RNA lies at the core of the protein synthesis machinery. Mitochondrial gene content is typically rather conserved in Metazoan (Boore *et al.*, 1995). With notable exceptions described in nematodes (Okimoto *et al.,* 1992), mollucs (Hoffmann *et al.,* 1992) and cnidarians (Beagley *et al.,* 1998). In general mitochondrial DNA (mtDNA) contains genes for two ribosomal sub units (12S and 16S), 22 tRNA, and 13 proteins (three subunits of cytochrome c oxidase, cytochrome B, subunits 6 and 8 of ATP F0 synthase, and seven subunits of NADH dehydrogenase). There is also a non-coding A+T rich region that contains signaling elements for regulation of replication and transcription (Wolstenholme, 1992). Animal mitochondrial genome size is usually about 16kb long, with few exceptions (Brown 1985). Gene rearrangements within the mtDNA, formerly considered rare, have been described for a number of taxa, including bees (Mindell *et al.,* 1998; Dowton and Campbell, 2001: Shao *et al.,* 2003; Silverstre *et al.,* 2002; and Silvestre and Arias, 2006).

MATERIALS AND METHODS

Insect Mitochondrial Genome: The complete mtDNA genome sequence from 11 different insect species representing 8 orders of class Insecta used in this study were retrieved from GOBASE the organelle genome database http://gobase.bcm.umontreal.ca and GenBank NCBI http://www.ncbi.nlm.nih.gov The dataset includes *Pyrocoelia rufa* (AF452048), *Triatoma dimidiata* (AF301594), *Campodea lubbocki* (NC008234), *Locusta migratoria* (NC001712), *Reticulitermes flavipes* (EF206314), *Adoxophyes honmai* (NC008141), *Anopheles gambiae* (L20934), *Anoplophora glabripennis*(DQ768215), *Antheraea pernyi* (NC004622) *Drosophila melanogaster* (NC001709*) Apis mellifera* (NC001566)which belongs to eight different orders (Diptera, Hymenoptera, Orthoptera, Coleoptera, Diplura, Lepidoptera, Isoptera, Hemiptera).

Phylogenetic Analysis: RNA alignment of all LSU rRNA and SSU rRNA sequences were performed using R-COFFEE (http://www.rcoffee.org) for Phylogentic and unrooted analysis followed by neighbor-joining method to construct phylogenies from sequence alignments using "PHYLIP" (Felenstein, 1989) and In order to assess the level of significance from the inferred clades, a bootstrap procedure was used following TREETOP - Phylogenetic Tree Prediction. http://www.genebee.msu.su/services/phtree_reduced.html.

RESULTS AND DISCUSSION

An unrooted phylogenetic tree showing the relationship between Diptera, Hymenoptera, Orthoptera, Coleoptera, Diplura, Lepidoptera, Isoptera, Hemiptera based on mitochondrial (large subunit)LSU and (Small subunit)SSU rRNA sequences were generated using

PHYLIP (Fig 1, 2, 3&4 respectively) Unrooted phylogenetic tree derived from mtDNA LSU rRNA and SSU rRNA sequences placed their insect species into taxonomic groups in separate orders similar to that of morphologically derived characteristics.

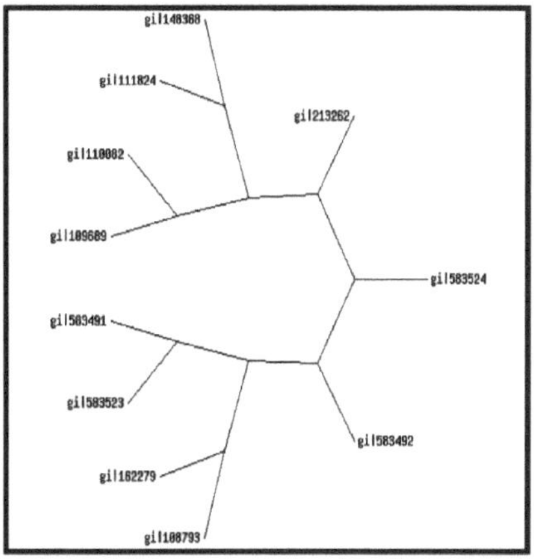

Fig. 1: LSU rRNA unrooted phylogenetic tree

The Phylogram of LSU rRNA revealed that both *Adoxophyes honmai, Antheraea pernyi* of Lepidoptera and *Anopheles gambiae, Drosophila melanogaster* of Diptera grouped under one sub cluster with supporting bootstrap value of 100 % while *Anoplophora glabripennis* (Coleoptera) and *Campodea lubbocki* (Diplura) grouped under same subcluster with supporting bootstrap value of 99%. *Triatoma dimidiate* (Hemiptera) and Reticulitermes *flavipes* (Isoptera) grouped under same cluster with supporting bootstrap value of 74% ,*Locusta migratoria* (Orthoptera) and *Apis mellifera* (hymenoptera) diverged with bootstrap value of 44 and 83 repectively. *Pyrocoelia rufa* (Coleoptera)divergedfrom the other clusters with supporting bootstrap value. The Phylogram of SSU rRNA revealed the grouping of *Adoxophyes honmai & Antheraea pernyi* of Lepidoptera under one subcluster and *Locusta migratoria* (Orthoptera)&*Anoplophora glabripennis* (Coleoptera) into another sub cluster with bootstrap value of 100. While *Anopheles gambiae* (Diptera), *Drosophila melanogaster* (Diptera) and *Apis mellifera* (hymenoptera) had bootstrap value of 63,14,13 repectively. *Reticulitermes flavipes* (Isoptera) and *Triatoma dimidiate* (Hemiptera) grouped under a subcluster with bootstrap value of 35. *Campodea lubbocki* (Diplura) diverged with bootstrap value of 31. *Pyrocoelia rufa* (Coleoptera) diverged from the other clusters with supporting bootstrap value of 100.

In LSU rRNA, cluster 1 show that insects belonging to Dipterans, Lepidopterans and Hymenopterans had genetic distance verymuch closer to Endopterygota. Orthopterans belonging to Exopterygota exist as a single cluster. In cluster 2, Coleopterans and Diplurans come under Endopterygota and Aptergota respectively.

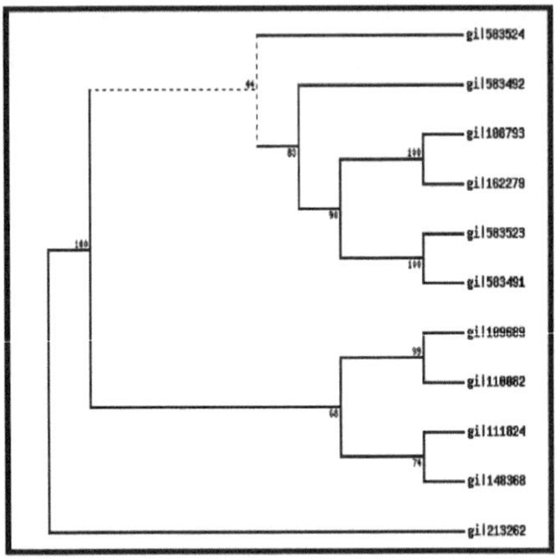

Fig 2: LSU rRNA phylogenetics tree

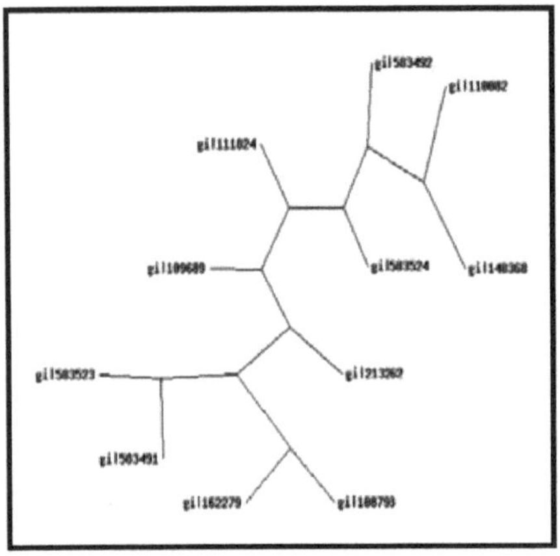

Fig. 3: SSU rRNA unrooted phylogenetic tree

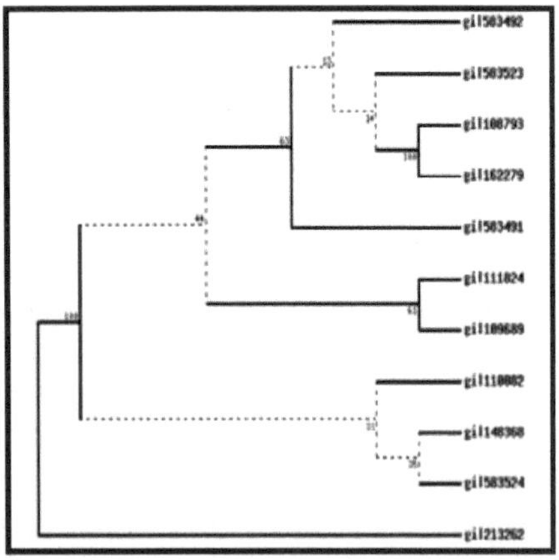

Fig. 4: SSU rRNA phylogenetics tree

Hemipterans and Isopterans of Exopterygota grouped into another cluster. Hymenopterans, Dipterans, Lepidopterans, Orthopterans and Coleopterans were grouped in cluster 1. Under this cluster, Orthoptera and Coleoptera grouped in a subcluster and Diplura, Coleoptera, Isoptera and Hemiptera were grouped in another subcluster. With current emphasis on identification of members of class Insecta based on mitochondrial genome, the new insights obtained will undoubtedly influence ideas on the origin of insect mitochondria.

REFERENCES

Beagley, C.T., Okimoto, R. and Wolstenholme, D.R. 1998. The Mitochondrial genome of the sea anemone Metridium senile (Cnidaria): Introns, a paucity of tRNA genes, and a near standard genetic code. *Genetics* 148:1091-1108.

Boore, J.L. 1999. Animal mitochondrial genomes. *Nucleic Acids. Res.* 27: 1767-1780.

Boore, J.L., Collins, T.M., Stanton, D., Daehler, L.L. and Brown, W.M. 1995. Deducing the patten of arthropod phylogeny from mitochondrial DNA rearrangements. *Nature*, 376:163-165.

Brown, W.M. 1985. The mitochondrial genome of animals. In: MacIntyre RJ (ed) *Molecular Evolutionary Genetics*. Plenus Press, New York, 95-130.

Felesenstein, J. 1989. PHYLIP Phylogeny Inference Package. *Cladistics*, 5: 164-166.

Hoffmann, R.J., Boore, J.L. 1992. A nowel mitochondrial genome organization for the blue mussel, *Mytilus edulis*. *Genetics*, 131:397-412.

Hudspeth, M.E.S. 1995. The fungal mitochondrial genome—a broader perspective. *In:* DK Aurora, PD Bridge and D Bhatnagar, eds. Handbook of Fungal Biotechnology. 1st ed. New York: Marcel Dekker, Inc. 213-241.

Mindell, D.P., Sorenson, M.D. and Dimcheff, D.E. 1998. Multiple independent origins of mitochondrial gene order in birds. *Proc Natl Acad Sci USA* 95:10693-10697.

Okimoto, R., Macfarlane, J.L., Clary, D.O. and Wolstenholme, D.R. 1992. The mitochondrial genome of two nematodes, *Caenorhabditis elegans and Ascaris Suum. Genetics*, 130:471-498 pp.

Palmer, J.D. 1990. Constrasting modes and tempos of genome evolution in land plant organelles. *Trends Genet.*, 6:115-120.

Shao, R., Dowton, M., Murrell, A. and Barker, S.C. 2003. Rates of gene rearrangement and nucleotide substitution are correlated in the mitochondrial genomes of insects. *Mol Biol Evol.*,20:1612-1619.

Silvestre, D., Francisco F.O., Weinlich, R. and Arias, M.C. 2002. A scientific note on mtDNA gene order rearrangements among highly eusocial bees. *Apidologie* 33:355-356.

Silverstre, D. and Arias, M.C. 2006. Mitochondrial tRNA gene translocations in highly eusocial bees. *Genet. Mol. Biol.*, 29:572-575.

5

Analysis of Microarray Gene Expression Data

M. Balakrishnan, R.C. Srivastava and M. Ramachandran

Two schemes of labeling are in common use today. One variant labels a single sample, either radioactively or fluorescently. Radioactive labeling is used, e.g., in conjunction with hybridization on nylon membranes (Lennon and Lehrach, 1991). The company Affymetrix synthesizes sets of short oligomers on a glass wafer and uses a single fluorescent label (Lipshutz *et al.*, 1999). Alternatively, two samples are labeled with a green and a red fluorescent dye, respectively. The mixture of the two mRNA preparations is then hybridized simultaneously to a common array on a glass slide. This technology is usually referred to as the Stanford technology (David Duggan *et al.*, 1999). Quantification utilizes a laser scanner that determines the intensities of each of the two labels over the entire array. Recently, companies like Agilent have immobilized long oligomers of 60 to 70 base pairs length and used two-color labeling.

The parallelism in this kind of experiment lies in the hybridization of mRNA extracted from a single sample to many genes simultaneously. The measured abundances, though, are not obtained on an absolute scale. This is because they depend on many hard to control factors such as the efficiencies of the various chemical reactions involved in the sample preparation, as well as on the amount of immobilized DNA available for hybridization.

There are many different ways for the outline of a microarray experiment. In many cases, a development in time is studied leading to a series of hybridizations following each other. Alternatively, different conditions like healthy/diseased or different disease types may be studied. We generally refer to a time point or a state as a condition and typically for each condition several replicate hybridizations are performed. The replicates should provide the information necessary to judge the significance of the conclusions one wishes to draw from the comparison of the different conditions. When going deeper into the subject it soon becomes clear that this simple outline constitutes a rather challenging program.

The goal of many microarray experiments is to identify genes that are differentially transcribed with respect to different biological conditions of cell cultures or tissue samples. Focuses on these issues, paying particular attention to the notoriously low numbers of repeated hybridizations per condition in relation to the high numbers of genes about which one wants to make conclusions. Section 5 proceeds to highlight some of the issues in pattern discovery in microarray data. Here, again, classical methods of data analysis need to be carefully evaluated with respect to their applicability to the particular type of data at hand. A short summary will be given of the methods that have so far been successfully applied. Emphasis is given to exploratory approaches that allow the subsequent formulation of hypothesis that can be tested either through further analysis or further experiments.

DATA VISUALIZATION AND QUALITY CONTROL

A microarray experiment consists of the following components: a set of *probes*, an *array* on which these probes are immobilized at specified locations, a *sample* containing a complex mixture of labeled biomolecules that can bind to the probes, and a *detector* that is able to measure the spatially resolved distribution of label after it has bound to the array (Yee Hwa Yang *et al*, 2002). The probes are chosen such that they bind to specific sample molecules; for DNA arrays, this is ensured by the high sequence-specificity of the hybridization reaction between complementary DNA strands. The array is typically a glass slide or a nylon membrane. The sample molecules may be labeled through the incorporation of radioactive markers, such as 33P- labeled probes, or of fluorescent dyes, such as phycoerythrin, Cy3, or Cy5.

After exposure of the array to the sample, the abundance of individual species of sample molecules can be quantified through the signal intensity at the matching probe sites. To facilitate direct comparison, the spotted array technology developed in Stanford (David Duggan *et al.,* 1999) involves the simultaneous hybridization of two samples labeled with different fluorescent dyes, and detection at the two corresponding wavelengths.

The total number of probes on an array may range from a few dozens to tens of thousands. Left panel: grey-scale representation of the detected labels fluorescence at 635 nm (red), corresponding to mRNA sample A. Right panel: label fluorescence at 532 nm (green), corresponding to mRNA sample B. Spots that light up in only one of the two images correspond to genes that are only transcribed in one of the two samples. Middle panel: false-color overlay image from the two intensity distributions. The spots are red, green, or yellow, depending on whether the gene is transcribed only in sample A, sample B, or both.

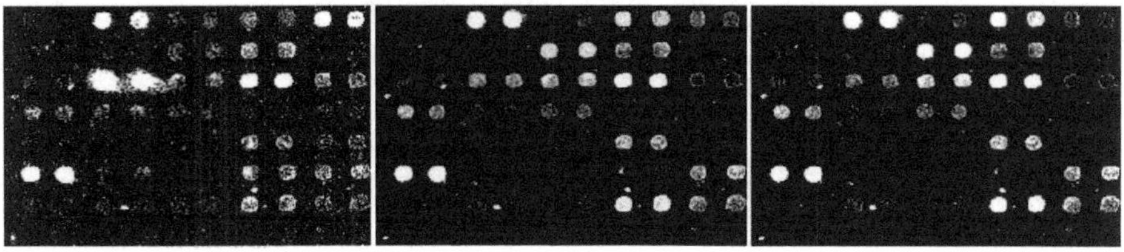

Fig 1: The detected intensity distributions from a cDNA microarray for a region comprising around 80 probes.

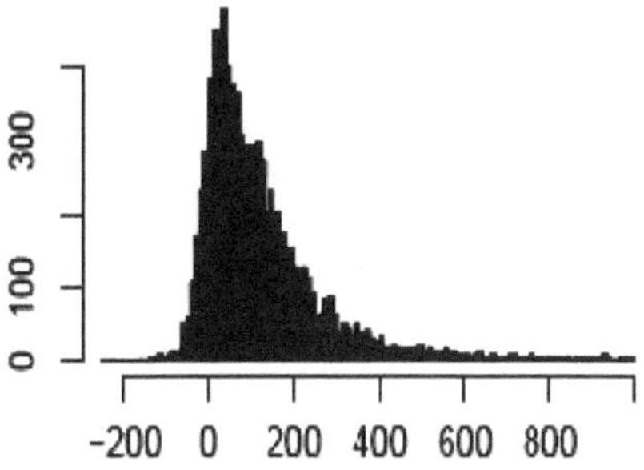

Fig 2: Histogram of probe intensities at the green wavelength for a cDNA microarray similar to the one depicted in Fig. 1.

IMAGE QUANTIFICATION

The intensity images are scanned by the detector at a high spatial resolution, such that each probe spot is represented by many pixels. In order to obtain a single overall intensity value for each probe, the corresponding pixels need to be identified (segmentation), and the intensities need to be summarized (quantification). In addition to the overall probe intensity, further auxiliary quantities may be calculated, such as an estimate of apparent unspecific "local background" intensity, or a spot quality measure. A variety of segmentation and quantification methods is implemented in available software packages. They differ in their robustness against irregularities and in the amount of human interaction that they require. Different types of irregularities may occur in different types of microarray technology, and a segmentation or quantification algorithm that is good for one platform is not necessarily suitable for another. For instance, the variation of spot shapes and positions that the segmentation has to deal with depends on the properties of the support (e. g. glass or nylon), on the probe delivery mechanism (e. g. quill-pen type, pin and ring systems, ink jetting), and on the detection method (optical or radioactive). Furthermore, larger variations in the spot positioning from array to array can be expected in home-made arrays than in mass produced ones. An evaluation of image analysis methods for spotted cDNA arrays was reported (Johannes Schuchhardt *et al*, 2000).For a microarray project, the image quantification marks the transition in the work flow from "wet lab" procedures to computational ones.

The intensities were determined, in arbitrary units, by an image quantification method, and "local background" intensities were subtracted. Due to measurement noise, this leads to non-positive probe intensities for part of the genes with low or zero abundance. The x-axis has been cut off at the 99% quartile of the distribution. The maximum value is about 4000. Number of data exploration and visualization methods that may be useful for these tasks.

DYNAMIC RANGE AND SPATIAL EFFECTS

A simple and fundamental property of the data is the dynamic range and the distribution of intensities. Since many experimental problems occur at the level of a whole array or the sample preparation, it is instructive to look at the histogram of intensities from each sample. An example is shown in Fig. 2. Typically, for arrays that contain quasi-random gene selections, one observes a unmoral distribution with most of its mass at small intensities, corresponding to genes that are not or only weakly transcribed in the sample, and a long tail to the right, corresponding to genes that are transcribed at various levels. In most cases, the occurrence of multiple peaks in the histogram indicates an experimental artifact. To get an overview over multiple arrays, it is instructive to look at the box plots of the intensities from each sample. Problematic arrays should be excluded from further analysis. There are different options for the intensity scaling, among them the linear, logarithmic, and rank scales. Each one will highlight different features of the spatial distribution. Examples are shown in Fig. 3.

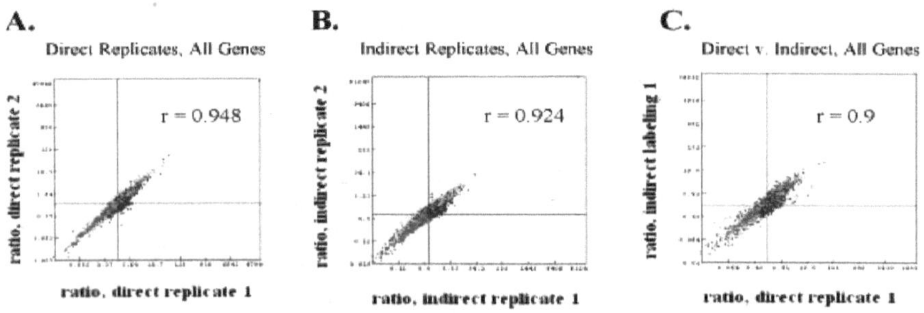

Fig. 3: Scatter plot images and corresponding Pearson Coefficients between direct and indirect labeling methods. (A) Scatter plot of direct labeling replicates yielding a Pearson Coefficient of 0.948 for all genes (n = 7,476 genes). (B) Scatter plot of indirect labeling replicates yielding a Pearson Coefficient of 0.924 for all genes (n = 7,474). (C) A representative image of a comparison of the gene expression ratios between a direct and indirectly labeled slide. The average Pearson Coefficient generated comparing labeling methods was 0.895 for all genes (range 0.883–0.911, n = 7,474 genes)

SCATTER PLOT

Usually, the samples hybridized to a series of arrays are biologically related, such that the transcription levels of a large fraction of genes are approximately the same across the samples. This can be expected e. g. for cell cultures exposed to different conditions or for cells from biopsies

The scatter plot allows assessing both measurement noise and systematic biases. Ideally, the data from the majority of the genes that are unchanged should lie on the bisector of the scatter plot. In reality, there are both systematic and random deviations from this (Sandrine Dudoit *et al*, 2002). For instance, if the label incorporation rate and photo efficiency of the red dye were systematically lower than that of the green dye by a factor of 0.75, the data would be expected not to lie on the bisector, but rather on the line y = 0:75x.

Rather than plotting a symbol for every data point, they use a density representation, which may be useful for larger arrays. For example, Fig. 4 shows the scatter plot from the comparison of two tissue samples based on 152,000 probes2. This uses the fact that the (1; 1; 1)-component of a three-way microarray measurement corresponds to average intensity, and hence is not directly informative with respect to differential transcription. Note that if the plotted data was pre-processed through a variance-stabilizing transformation, its variance does not depend on the (1; 1; 1)-component.

BATCH EFFECTS

Present day microarray technology measures abundances only in terms of relative probe intensities, and generally provides no calibration to absolute physical units. Hence, the comparison of measurements between different studies is difficult. Moreover, even within a single study, the measurements are highly susceptible to *batch effects*. By this term, we refer to experimental factors that (i) add systematic biases to the measurements, and (ii) may vary between different subsets or stages of an experiment.

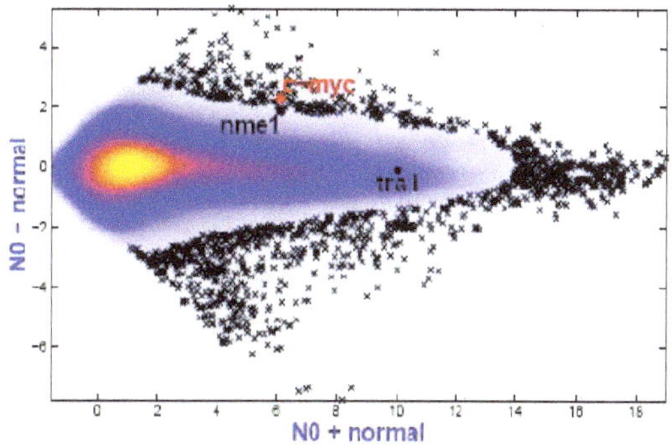

Fig 4: Scatterplot of a pairwise comparison of non-cancerous colon tissue and a colorectal tumor.

Individual transcripts are represented by 'x' symbols. The x-coordinate is the average of the appropriately calibrated and transformed intensities. The y-coordinate is their difference, and is a measure of differential transcription. The array used in this experiment contained 152,000 probes representing around 70,000 different clones. Since plotting all of these would lead to an uninformative solid black blob in the centre of the plot, the point density is visualized by a color scale, and only 1500 data points in sparser regions are individually plotted.

Spotting: To manufacture spotted microarrays, the probe DNA is deposited on the surface through spotting pins. Furthermore, the efficiency of a pin may change over time through mechanical wear, and the quality of the spotting process as a whole may be different at different times, due to varying temperature and humidity conditions.

PCR amplification: For cDNA arrays, the probes are synthesized through PCR, whose yield varies from instance to instance. Typically, the reactions are carried out in parallel in 384-well plates, and probes that have been synthesized in the same plate tend to have highly correlated variations in concentration and quality.

Sample preparation protocols: The reverse transcription and the labeling are complex biochemical reactions, whose efficiencies are variable and may depend sensitively on a number of hard-to-control circumstances. Furthermore, RNA can quickly degrade; hence the outcome of the experiment can depend sensitively on when and how conditions that prevent RNA degradation are applied to the tissue samples.

Array coating: Both the efficiency of the probe fixation on the array, as well as the amount of unspecific background fluorescence strongly depends on the array coating.

Scanner and image analysis: Top panel: scatter plot of intensities from a pair of cDNA arrays, comparing renal cell carcinoma to matched non-cancerous kidney tissue. Similar to Fig. 4, the x-coordinate represents average, and the y-coordinate differential signal. In the bottom of the plot, there is a cloud of probes that appear to represent a cluster of strongly down-regulated genes. Since the clone selection was quasi-random, this points to a defect in the probe synthesis that affected one array, but not the other. The discovery of such artifacts may be helped by coloring the dots in the scatter plot by attributes such as PCR plate of origin or spotting pin (for technical reasons, the print version of this figure is shown in grey-scale). While the example presented here is an extreme one, caution towards batch artifacts is warranted whenever arrays from different manufacturing lots are used in a single study.

These considerations have important consequences for the experimental design: first, any variation that can at any means be avoided within an experiment should be avoided. Second, any variation that cannot be avoided should be organized in such a manner that it does not confound the biological question of interest. Clearly, when looking for differences between two tumor types, it would not be wise to have samples of one tumor type processed by one laboratory, and samples of the other type by another laboratory.

LIMITATIONS

The concepts of Section 3.1 have been widely used for microarray data analysis. However, it has also become clear that, for many data sets that are encountered in practice, they are not sufficient. The following points are worth noting.

Robustness: In order to make model (David Duggan *et al.*, 1999) identifiable, Chen *et al.,* assumed that the transcription levels of all genes were unchanged, and set _k1 = _k2 for all k. Thus, their model is misspecified for the part of the data arising from truely differentially transcribed genes, which act as outliers. However, their parameter estimation is based on least- squarescriteria, and may be sensitive to the presence of such outliers. Besides that, outliers may be caused by technical artifacts.

Heteroskedasticity: The significance of log-ratios depends on the absolute values of the intensities in the numerator and denominator. Typically, the variance of log-transformed intensities increases as their mean decreases.

Apparent non-linearities: According to the above models, the data from a pair of samples should lie along a straight line in the scatterplot of the log-transformed intensities. However, in real data, several authors have observed data that follows a curved line.

Negative values: While the image quantification's estimates for probe "foreground" and "background" intensities are generally positive, this is usually not true for their difference. If a gene is weakly or not expressed, it can happen by chance that the background estimate is larger than the foreground estimate. However, non-positive values make sense neither for ratios nor for the log-transformation.

IDENTIFICATION OF DIFFERENTIALLY EXPRESSED GENES

One of the basic goals in the analysis of microarray gene expression data is the identification of differentially expressed genes in the comparison of different types of cell or tissue samples. In order to control the biological and experimental variability of the measurements, statistical inference has to be based on an adequate number of replicate experiments. Here one may distinguish between cases in which one wishes to make statements on a particular cell population and cases in which one wants to make statements that hold in the presence of biological variability, such as with biopsy studies of diseased tissues. In the first case, independent replications can be obtained on the level of multiple mRNA isolations, in the second, they may be provided by samples from different patients.

To identify differentially expressed genes with respect to a certain biological question, a suitable statistical test may be performed for each gene. The choice of the test statistic depends on the biological question and on the nature of the available experimental data. In the simplest case, one asks for genes that show different transcript abundance between two conditions. In more complex situations, one may look for genes whose abundance is associated with multiple factor levels of one or more sample characteristics. Furthermore, one may consider continuous-valued sample characteristics and test for genes which show non-zero coefficients in a regression model, such as a linear model or a Cox proportional hazards model on patient survival data.

Different statistical tests may make more or less strong assumptions on the distributions of replicate measurements. Important questions are whether the distributions are symmetric, how similar or dissimilar they are to normal distributions, what their behavior at the tails is like, and whether or how their variance (or another appropriate measure of scale) varies between different genes or between different conditions). There may be other technological effects that can influence the variance of the measurement distributions in a gene- or condition-dependent manner, such as GC-content, or probe length. Finally, there may be genuine biological differences, such as different tightness of the regulatory control for different genes or for the same gene under different physiological or disease conditions.

Unsupervised as well as supervised learning methods play a central role in the analysis of microarray gene expression data. Supervised methods aim at inferring information from the data with respect to a pre-defined response variable. For instance, in the context of tumor diagnostics one tries to classify mRNA samples obtained from tumor cells with respect to given tumor types. In the following, we focus on unsupervised methods, which aim at detecting structures in the data without making use of gene or sample annotations. A primary purpose of such methods is to provide a visualization of the data in which conspicuous structures can easily be recognized. These may be relations among genes, among samples, or between genes and samples. The perception of such structures can lead the researcher to develop new hypotheses: e.g. the result of a clustering of genes may indicate the putative involvement of uncharacterized genes in a biological process of interest, whereas a separation of the expression profiles of a set of patient tissue samples into clusters may point to a possible refinement of disease taxonomy. On the other hand, unsupervised methods are often used to confirm known differences between genes or samples on the level of gene expression: If a clustering algorithm groups samples from e.g. two different tumor types into distinct clusters without using prior knowledge, this provides evidence that the tumor types do indeed show clearly detectable differences in their global gene expression profiles. For all of the following methods, we assume that we have a gene expression data matrix of suitably calibrated and transformed expression levels with, say, the rows corresponding to genes and the columns corresponding to cell or tissue samples.

CLUSTER ALGORITHMS

Cluster algorithms generally aim at grouping objects according to some notion of similarity. An overview of questions and methods in cluster analysis is given in. For microarray data, clustering may be applied to the genes whose expression levels are measured, with the expectation that functionally related or co–regulated genes will show similar expression patterns. On the other hand, one may use clustering to analyze the expression profiles of a set of cell or tissue samples with the hope that samples with similar biological characteristics will be grouped together. Cluster algorithms are explicitly or implicitly based on a quantitative measure of dissimilarity between the objects of interest. In the case of row and column vectors of a gene expression data matrix, typical examples are the Euclidean distance or 1 minus the correlation coefficient. For the clustering of genes based on their expression patterns, the latter is often used because it is invariant under affine–linear transformations of the input vectors and focuses on the pattern of relative changes.

In *hierarchical* clustering algorithms, a tree structure (dendrogram) is computed that contains the objects as leaves. *Agglomerative* methods build the tree starting with the leaves, whereas in *divisive* methods, the set of objects is iteratively partitioned into subsets. To obtain a partition of the set of objects into clusters, the resulting dendrogram may be cut at a certain height. In the analysis of microarray data however, hierarchical clustering is often simply used to obtain a linear ordering of both the rows and the columns of a gene expression data matrix such that similar rows or columns are located close to each other. The reordered data matrix is then displayed using a color map, which may be a powerful visualization tool. An important, but difficult question in cluster analysis is that of the validity of the results. Cluster analysis assumes that the data are organized in distinguishable clusters. However, a cluster algorithm will usually also produce a set of clusters if this assumption is not fulfilled. Furthermore, the results may be affected by random fluctuations of the data. Thus, it is of interest to estimate the number of clusters present in a data set (if any), as well as to assess the variability of various features of the result.

The performance of this approach was demonstrated using simulated data and real gene expression data, where clustering is applied to the samples. However, one might imagine that real data sometimes lie in–between the extreme cases of a common distribution for all objects on one side and distinct clusters on the other side. Furthermore, if the objects are organized in hierarchically nested clusters, such that there are partitions at different levels of granularity, the question of how many clusters are present is not meaningful without further specifications. Kerr and Churchill used the bootstrap in order to assess the reliability of the results of a cluster analysis. Resampling was performed on the residuals of an analysis of variance model, yielding stability values for the assignment of genes to pre-specified clusters. In a more general context, Pollard and van der Laan analysed the performance of the bootstrap in assessing the variability of the results of a wide class of clustering methods. Assuming that the expression profiles of the biological samples are generated from a mixture of n-variate probability distributions, where n denotes the number of genes, they gave a general definition of clustering methods (for clustering genes or samples, or the simultaneous clustering of genes and samples) as algorithms to estimate certain parameters of the data-generating distribution. Although an underlying probabilistic model is assumed, the cluster algorithms under consideration do not have to be model-based. This framework allows applying concepts of statistical inference to clustering algorithms. In a simulation study, they analyzed both the non-parametric resampling of sample expression profiles and the parametric bootstrap based on normal distributions. The results indicated that both bootstrapping methods are able to assess the variability of various quantities describing the output of a clustering algorithm. In addition, the authors proposed to test the statistical significance of a clustering of samples via the comparison with data generated from an appropriate null model. A similar approach for the evaluation of temporal gene expression patterns was presented in. Assuming the expression pattern of each gene to be generated from one of several multivariate normal distributions, the authors analyzed the error rate of various clustering algorithms in determining the correct cluster membership of genes.

CONCLUSION

We have described different aspects of microarray gene expression data analysis, from the quality control of the raw probe intensity data, via calibration, error modeling, and the identification of differentially transcribed genes to explorative methods such as clustering or pattern discovery. Yet, statistical analysis is only one part of a microarray experiment. Frequently, data analysis is expected to correct for technological problems or shortcomings in the design of an experiment. Awareness has grown, though, in recent years and interaction between experimentalists and data analysts is improving and, very importantly, starting already early in the planning of an experiment. This raises the hope that statistical analysis will in the future be ever less diverted by trouble shooting, and can increase its focus on generating and validating biological hypotheses from the data. Microarray based experiments are frequently seen as the stronghold of hypothesis-free genome research. While debatable in itself, this assertion simply shifts the responsibility to the computational scientist analyzing the data. In the absence of a clear hypothesis much of the analysis will be of an exploratory nature. Once this leads to a hypothesis further independent verification is needed. This embeds microarray experiments and statistical analysis into a feedback cycle producing new experiments

REFERENCES

Lennon, G.G. and Lehrach, H. 1991. Hybridization analyses of arrayed cDNA libraries. *Trends in Genetics*, 10:314-317.

Lipshutz, R.J., Fodor, S.P., Gingeras, T.R., and Lockhart, D.J. 1999. High density synthetic oligonucleotide arrays. *Nature Genetics*, 21 (Suppl 1):20-24.

David, J. Duggan, Michael Bittner, Yidong Chen, Paul Meltzer, and Jeffrey, M. Trent. 1999. Expression profiling using cDNA microarrays. *Nature Genetics*, 21 (Suppl 1):10–14 pp.

Yee Hwa Yang and Terence, P. Speed. 2002. Design issues for cDNA microarray experiments. *Nat. Rev. Gen.*, 3:579–588.

Yee Hwa Yang, Buckley MJ, Sandrine Dudoit and Terence P Speed. 2002. Comparison of methods for image analysis on cDNA microarray data. *Journal of Computational and Graphical Statistics*, 11:108–136.

Johannes Schuchhardt, Dieter Beule, Arif Malik, Eryc Wolski, Holger Eickhoff, Hans Lehrach, and Hanspeter Herzel. 2000. Normalization strategies for cDNA microarrays. *Nucleic Acids Research*, 28:e47.

Sandrine Dudoit, Yee Hwa Yang, Terence P. Speed, and Matthew J. Callow. 2002. Statistical methods for identifying differentially expressed genes in replicated cDNA microarray experiments. *Statistica Sinica*, 12:111-139.

6

A Novel Sequence Encoding Method for Searching Rareness Among Consensus DNA Sequences

K. Meena, K. Menaka, T.V. Sundar and K. R. Subramanian

In the context of molecular biology, a consensus sequence is the minimum nucleotide sequence that shows the characteristics common to most members of the family among a set of related genetic sequences.It serves in representing the results of a multiple sequence alignment, where related sequences are compared to each other, and similar functional sequence motifs are found. It also shows conserved/variable residues. It is found several times in the genome and is thought to play the same role in its different locations. For example, many transcription factors recognize particular consensus sequences in the promoters of the genes they regulate. In the same way restriction enzymes usually have palindromic consensus sequences, usually corresponding to the site where they cut the DNA. Transposons act in much the same manner in their identification of target sequences for transposition. Splice sites, *i.e.* those sequences immediately surrounding the exon-intron boundaries, can also be considered as consensus sequences. Defining a consensus DNA binding site by examining consensus sequences is a key work carried out by biologists (Harley *et al.*, 1994). Rareness or anomaly detection followed by in depth analyses in such sequences may reveal significant information regarding the structural, functional and biochemical mechanisms of the genes which are of much importance to biologists.

ENCODING THE SEQUENCES

Most of the sequence encoding techniques reported in the literature for mathematical manipulation of sequences for studies and analyses are associated with constraint based methods such as gap penalties, ascribing weights, matrixing the observed features *etc*. As far as the knowledge of the authors is concerned, there is no direct and non-constraint encoding mechanism of biological sequences reported in the literature so far. In this context, the authors (Meena *et al.,* 2008) suggest a new method of encoding, simple and direct, for transforming the DNA sequences into numerical entities for mathematical and statistical studies. The method is just to transform the nucleotide bases into their corresponding ASCII (American Standard Code for Information Interchange) values. In this work, the bases present in the given nucleotide sequence are converted to numerical equivalents (65 for A, 71 for G, 67 for C and 84 for T) that correspond to the upper case letters of them. After carrying out this process, the resulting sequence will become the numerical analogue of the given original DNA sequence. The numerical encoding may also be done with ASCII values corresponding to lower case letters. This directly encoded numerical sequence may subsequently be subjected to mathematical and statistical analyses.

METHODOLOGY

The sequences to be studied are first taken in FASTA format (Pearson, 2000). The individual sequences are readied in separate text files, after deleting the first comment line occurring in the FASTA format. The screen shot of the selected sequences (in the text boxes) and the program output is shown in Figure 1.

ANALYSIS OF VARIANCE

The statistical method of Aalysis Of VAriance (ANOVA) is used to determine the presence or absence of variation among a group of numerical data sets. It is one of the most elegant, powerful and useful techniques for studying total variation in a set of data which may be reduced to components associated with possible scores of variability whose relative importance is needed. It involves the calculation of variance values of every member of the group elements from the mean value of that group and subsequently analyzing the total variation between inter (variation between groups) and intra (variation within a particular group's elements) sequences under analysis. In the present method, the encoded numerical analogues of the DNA sequences serve as the data sets for the ANOVA study. The steps involved in the ANOVA procedure are given in the following algorithm.

ANOVA Algorithm

Step 1: Identification of the number of sequences for analysis. At present the program can handle a maximum of eight sequences at a time with individual sequence having lengths not exceeding 5000 base pairs.

Step 2:Computation of the Mean Values of the sequences (by using the numerically encoded sequences).

$$\text{Total Variation (TV)} = \sum_{i=1}^{N} (A_i - \hat{A})^2 = \sum A_i^2 - (\sum A_i)^2 / N$$

where $\hat{A}_s$ = Mean of the Sequence and $\hat{A}$ = Grand Mean

Step 3:Calculation of variances for the sequential elements using the mean value of the sequence and total variation.

Step 4:Calculation of Variation between Sequences (VBS: inter variance) and Variation within Sequences (VWS: intra variance).

$$\text{VBS} = \sum_{i=1}^{K_s} N_s (\hat{A}_s - \hat{A})^2$$

Where $\hat{A}_s$ = Mean of the Sequence, $\hat{A}$ = Grand Mean and

N_s = Number of Items in the sample sequences

$$\text{VWS} = \sum_{i=1}^{K_s N_s} {}_{Ai^2} - \sum_{i=1} [(\sum_{Ai})^2] / N_s$$

Where $\sum_{i=1}^{K_s}$=Summation over K_s Sequences (or)

VWS = Total Variation (TV) – Variation between Sequences (VBS)

Step 5: Calculation of Degrees of Freedom (DOF) for VBS (V1) and VWS (V2).

Step6: Estimation of the variances for VBS (S1) and VWS (S2).

DOF for VBS (V1) = Number of Sequences -1

DOF for VWS (V2) = Number of Items in all Sequences – Number of Sequences

S1 = Sequence sum in VBS / V1

S2 = Squared sum in VWS/ V2

F-ratio = Estimation of VBS / Estimation of VWS = S1/S2

Step 7: Computation of the statistically calculated F-ratio and comparison of this value with table F-ratio.

Step 8: Comparison with table F-ratio and deduction of result:

If Table F-ratio < Computed F-ratio, then there is a scope for the anomaly among the studied sequences. Otherwise, it is the indication for the existence of consensus.

Step 9: Interpretation of Results:

The table value of F is compared with the calculated value of F for testing the validity of null hypothesis i.e. the means among two or more groups are equal, under the assumption that the sampled populations are normally distributed. When the null hypothesis of equal means is true, the two mean squares estimate the same quantity (error variance), and should be of approximately equal magnitude. In other words, their ratio should be close to 1. If the null hypothesis is false, VBS should be larger than VWS. The table values of F are available at various degrees of freedom for different levels of significance. Usually at five percent level the table F- values (Engineering Statistics Handbook, 2008) are taken and so presented in this work also.

RESULTS AND DISCUSSION

The verification for the existence of consensus or rareness is judged based upon the calculation of the variation within the sequence data and the group of sequences. For an illustration, the authors used some DNA consensus sequences corresponding to some rodent Alu repetitive elements submitted in Repbase (Jurka *et al.,*, 2008: Repbase version 13.08) and presented the observations and results in Figure 1. The Alu elements are short interspersed elements (SINEs) ~300 nucleotides in length. Recent findings suggest that Alu elements may have a broad evolutionary impact by affecting gene structures (Dagan *et al,*, 2004), protein sequences, splicing motifs and expression patterns. Presently, such data are limited since the structural and positional information on genes and Alu sequences are scattered throughout incompatible and unconnected databases. Use of verification tools like the present method may serve in achieving the confluence of relevant data.

For illustration, sequence data viz. short interspersed repeats in rodents are used (Figure 1). It can be observed from the table that there is a complete matching between the conclusions suggested by this method and the corresponding Repbase documentation for the interspersed repeats. In all the studied cases the calculated F-ratios are very much lower than the Figure F-ratios, thus indicating the consensus of sequences.

In the cases of exception, the calculated F-ratios are expected to be slightly higher than the Figure F-ratios, reporting anomaly among sequences. The authors are working on such cases also and the detailed results will be published elsewhere. Further the closeness in the F-ratio values in exceptional cases, indicate the result in favour of consensus.

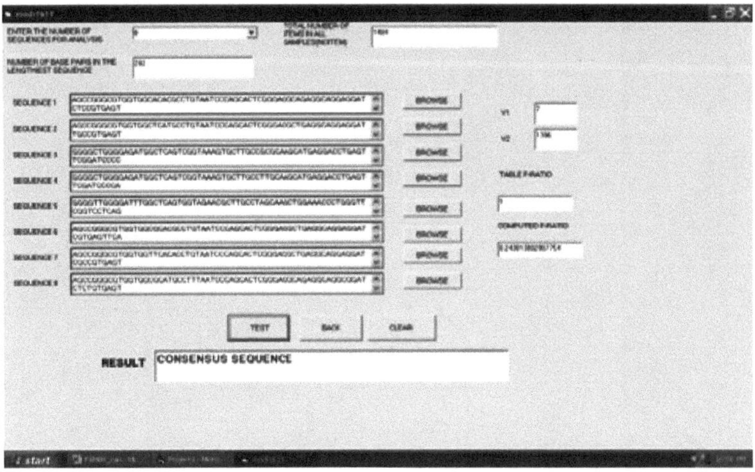

Fig. 1: A Screen shot view of the program

CONCLUSION

Statistical approaches like the above method, though seems to be rudimentary are effective when combined with experimental work and may serve in enhancing the precision of computational identification of potential consensus sequences in genomic DNA. Use of such computational verification tools not only saves precious time for biologists but also give them a helping hand in choosing closely related data from the multitude of genomic data. Finally, the suggested sequence encoding mechanism and related statistical methods can be validated by applying to other sequence related problems and may be tried in training neural nets also.

REFERENCES

Engineering Statistics Handbook. 2008.http://www.itl.nist.gov/div89 8/handbook

Jurka, J., Kapitonov, V.V., Pavlicek, A., Klonowski, P., Kohany, O., Walichiewicz, J. 2005. Repbase Update, a database of eukaryotic repetitive elements. *Cytogentic and Genome Research*, 110:462-467.

Meena, K., Menaka, K., Sundar, T.V. and Subramanian, K.R. 2008. [Communicated].

Pearson WR, 2000. Flexible sequence similarity searching with the FASTA3 program package. *Methods Mol. Bio.* 132:185-219.

Repbase version 13.08. 2008. FASTA and EMBL formats. GIRI: Genetic Information Research Institute. www.Girinst.orgTal Dagan, Rotem Sorek, Eilon Sharon, Gil Ast and Dan Graur. 2004. *Nucleic Acids Research*, (32): D489-D492.

Vincent, R. Harley, Robin Lovell-Badge and Peter, N. Goodfellow. 1994. *Nucleic Acids Research*, 22(8): 1500-1501.

7

Statistical Methods for Genomic Sequence and Microarray Analysis

A.K. Roy and S.R. Martha

ANALYSIS OF CODON USAGE VARIATION

Multivariate analysis methods such as Correspondence Analysis and Principle Component Analysis have often been used to identify major trends of variation in synonymous codon usage among inter or intra specific genes. Genetic codes are degenerate meaning most of amino acids can be encoded by more than one codon (triplet of nucleotides); such codons are synonymous and usually differ by one nucleotide in the third position. The alternative synonymous codons are not used with similar frequency and their usage varies among different genes. Hence there is the presence of Relative Synonymous Codon Usage values.

SEQUENCE DATA

The gene sequences downloaded from Genbank of National Center for Biotechnology Information that is the nucleotide sequence repository known worldwide, serves as the secondary data used for analysis. Relative synonymous codon usage value of each gene is extracted separately using the Codon Usage software of Sequence Manipulation Suite. These are the raw data which needs to be normalized to be used before application of various multivariate analysis.

RELATIVE SYNONYMOUS CODON USAGE

The first stage normalized data are absolute codon frequencies to avoid biases, such as gene length and amino acid usage, which may mask variation in synonymous codon usage. The effect of gene length can be avoided by dividing the frequencies of codons in a gene by the total codon frequency in the same gene, but such normalized data can be affected by amino acid usage in the sequence. To avoid the effect of amino acid usage, the codon frequencies are often normalized for each individual amino acid. However, two such sets of frequently used normalized data, relative codon frequencies (the number of times the codon is observed i.e the frequency of occurrence of the codon in a gene) and relative synonymous codon usage. RSCU values are a reflection of how often a particular codon is used relative to the expected number of times that codon would be used in the absence of codon usage bias. It is the relative frequencies of synonymous codons to code for same 0.amino acid. Synonymous codons form the subset of codons coding for a single amino acid.

CORRESPONDENCE ANALYSIS

Correspondence Analysis is a multivariate exploratory graphical technique used to describe the relationship between the row and column variables of a contingency table. A matrix is constructed keeping genes as row variable and codons as column variable and RSCU values as weight variables that is uploaded into the statistical package SPSS to carry out Correspondence Analysis.

CONSTRUCTION OF MATRIX

The result generated in form of many tables and plots. It gives output in form of correspondence table, table of dimensionality, profiles and distances, row and column scores, eigen values, mass, centroid, singular values, confidence statistics, plots of symmetrical normalization. A correspondence table is any two-way table whose cells contain some measurement of correspondence between the rows and the columns. The measure of correspondence can be any indication of the similarity, affinity, confusion, association, or interaction between the row and column variables.

Table 1: Hypothetical table with genes forming rows and codon representing columns

Codons	Codan1	Codan2	Codan3	...	Codan62	Codan63	Codan64
Genes							
Gene 1	$RSCUg_1c_1$	$RSCUg_1c_2$	$RSCUg_1c_3$		$RSCUg_1c_{62}$	$RSCUg_1c_{63}$	$RSCUg_1c_{64}$
Gene 2	$RSCUg_2c_1$	$RSCUg_2c_2$	$RSCUg_2c_3$		$RSCUg_2c_{62}$	$RSCUg_2c_{63}$	$RSCUg_2c_{64}$
Gene 3	$RSCUg_3c_1$	$RSCUg_3c_2$	$RSCUg_3c_3$		$RSCUg_3c_{62}$	$RSCUg_3c_{63}$	$RSCUg_3c_{64}$
...							
Gene 98	$RSCUg_{98}c_1$	$RSCUg_{98}c_2$	$RSCUg_{98}c_3$		$RSCUg_{98}c_{62}$	$RSCUg_{98}c_{63}$	$SCUg_{98}c_{64}$
Gene 99	$RSCUg_{99}c_1$	$RSCUg_{99}c_2$	$RSCUg_{99}c_3$		$RSCUg_{99}c_{62}$	$RSCUg_{99}c_{63}$	$SCUg_{99}c_{64}$
Gene 100	$RSCUg_{100}c_1$	$RSCUg_{100}c_2$	$RSCUg_{100}c_3$		$RSCUg_{100}c_{62}$	$RSCUg_{100}c_{63}$	$SCUg_{100}c_{64}$

A correspondence table is a cross tabulation, where the cells contain frequency counts. Correspondence analysis computes the row and column profiles to determine the distance between categories. Row profiles indicate the proportion of the row category in each column category. It corresponds to the cell contents divided by their corresponding row total. Similarly column profiles indicate the proportion of the column in each row category. Likewise, they are the elements divided by the column marginal. Mass is a measure that indicates the influence of an object based on its marginal frequency.Point with lagre mass pulls the centroid strongly to their location. Points with small mass pull the row centroid only slightly to its location. Correspondence analysis solution represents the relationship between row and column variables in as few dimensions as possible. It is frequently useful to look at the maximum number of dimensions to see the relative contribution of each dimension. Eigen values (inertia) are the percent of variance explained by each dimension. The row and column scores are the coordinates of the row and column points in the correspondence map.

CORRESPONDENCE MAP

Correspondence Analysis generates a variety of plots that graphically illustrate the underlying relationships between categories and between variables. Biplot is the scatter plot of the row and column scores for the two-dimensional solution.

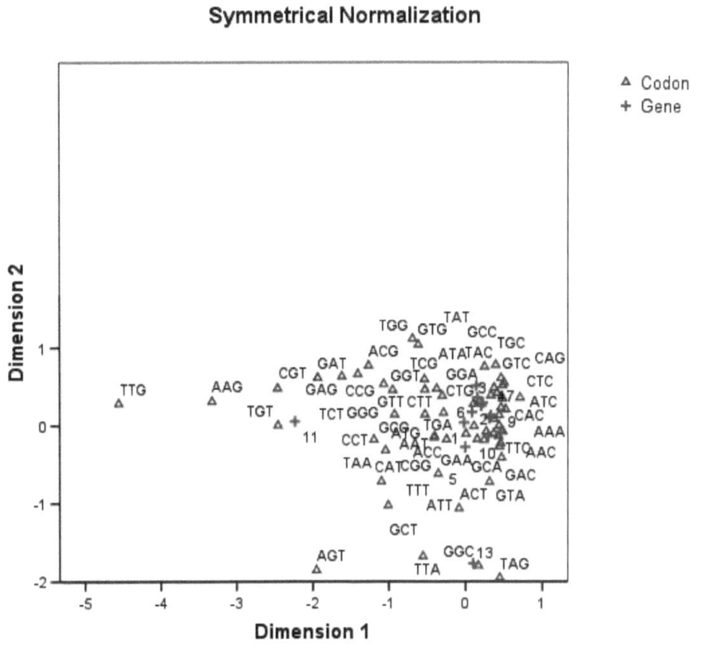

Fig. 1: Biplot between genes and codons of *C. carpio* generated by Correspondence Analysis of SPSS

In the example given in this paper, a plot between 13 genes and 64 codons of *Cyprinus carpio* is displayed. The plot reveals that except two genes that are located far away from the origin, all other genes follow a common trend. The genes 11 is associated with codons TGT, AAG, CGT and gene 13 is associated with codons GGC, TTA, TAG while all other genes viz. 1, 2, 3, 4, 5, 6, 7, 8, 9, 10 and 12 prefer GCG, CGG, TGC, TAC, ATA, TCG, GGA, GTC, CAG, CTC, CTG, CTT, ATC, CAC, TGA, ATG, AAT, ACC, GAA, TTC, AAC, GCA, GAC, ACT, GTA etc over other codons for synthesizing their respective protein. It is clear from this that the genes are baised to GC3 content.

PRINCIPLE COMPONENT ANALYSIS

Principal components analysis can be thought of as a method of dimension reduction. A set of variables is analyzed to reveal major dimensions of variation. The original data set can then be replaced by a new, smaller data set with minimal loss of information. The method reveals relationships among variables, among cases, and among variables and cases. The criterion used by categorical principal components analysis for quantifying the observed data is that the object scores (component scores) should have large correlations with each of the quantified variables.

Glancing at the quantifications in the joint plot of the category points, you can see that some of the categories of some variables were not clearly separated by the categorical principal components analysis as cleanly as would have been expected

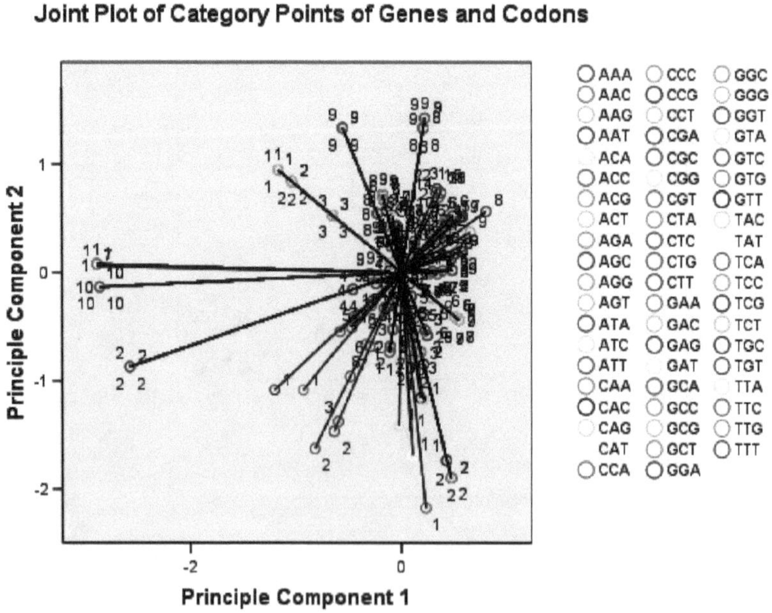

Fig. 2: An example of Principle component Analysis generated by SPSS.

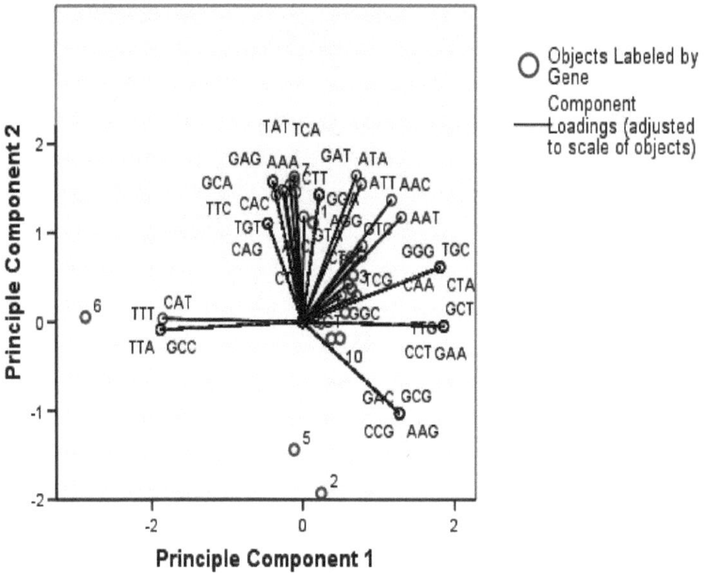

Fig. 3: An example of Principle component Analysis generated by SPSS

if the level had been truly ordinal. The joint plot of category points resembles the plot for the component loadings, but it also shows where the endpoints are located that correspond to the lowest quantifications. Figure 2 displays an example of categorical principle component analysis carried out between genes and codons using codons as ordinal variables.

MICROARRAY ANALYSIS

Microarray provides insight into the transcriptional state of the cell, measuring RNA levels for thousands of genes at once. It requires preparation of mRNA from cells growing under certain experimental conditions. Subsequently, the labeled cDNA mixture is hybridized to the microarray. After detection of the signals, image analysis programs are used to determine spot intensities. Data produced may be regarded as a table, each row representing a gene, each column standing for an experimental condition. Multiple measurements for each condition, involving repeated sampling, labeling, and hybridization, offer the opportunity of extracting more robust signals. The columns of the table have to undergo a normalization procedure, correcting for affine-linear transformation among the columns. All of the methods recently used for microarray data analysis would result in an outline of applied statistics, however. Most methods fall into one of three groups, namely clustering, classification, and projection methods. Examples of clustering techniques are k-means clustering, hierarchical clustering, and self-organizing maps. Classification methods take as input a grouping of objects and aim at delineating characteristic features common and discriminative to the objects in the groups. Examples of classification methods range from linear discriminate analysis to support vector machines or classification and regression trees.

GENERATION OF MICROARRAY IMAGES

There are a variety of microarray platforms that have been developed to accomplish and the basic idea for each is a glass slide or membrane that is spotted or "arrayed" with DNA fragments or oligonucleotides that represent specific gene coding regions. Purified RNA is then fluorescently- or radioactively labeled and hybridized to the slide/membrane. In some cases, hybridization is done simultaneously with reference RNA to facilitate comparison of data across multiple experiments. After thorough washing, the raw data is obtained by laser scanning or auto radiographic imaging. The final step of the laboratory process is to produce an image of the surface of the hybridized array. Fluorescently labeled microarrays can then be "read" with commercially available scanners scanning confocal microscopes with lasers exciting at wavelengths specifically for Cy3 and Cy5. The amount of signal emitted is directly in proportion to the amount of dye at the spot on the microarray and these values are obtained and quatitated on the scanner. This results in the image that is typical microarray picture.

PREPROCESSING

Preprocessing of microarray data comprises analytical or transformational procedures that are directed at resolving the systematic error and bias.

1. **Checking the background Information:** Background and foreground intensities together form the spot intensity. If the spot intensities are dependent on the background intensities, it is possible either not to apply any background correction to the data or discard the deviating observations from further analyses. The background corrected intensity values are calculated spot wise by subtracting the background intensity from the spot (foreground) intensity.

2. **Calculation of expression change:** There are three commonly used measures of expression change. Intensity ratio is the raw expression value, and log ratio and fold change are transformationally derived from it.

   i) **Intensity ratio:**After background correction, expression change is calculated. The simplest approach is to divide the intensity of a gene in the sample by the intensity level of the same gene in the control. Intensity ratio can be calculated from background correctedor uncorrected data (here we use corrected data) The formula for two-color data isFor affymetrix data, substitute Cy3' and Cy5' with the appropriate intensitiesfrom the sample and control chips. This intensity ratio is one for an unchanged expression, less than one for down regulated genes and larger than one for up-regulated genes. The problem with intensity ratio is that its distribution is highly asymmetric or skewed. Up-regulated genes can take any values from one to infinity (1 to ¥) whereas all down regulated genes are squeezed between zero and one (0 to 1). Such distributions are not very useful for statistical testing.

$$\text{Intensity ratio} = \frac{Cy3'}{Cy5'} \qquad \text{Intensity ratio} = \frac{Cy3'}{Cy5'}$$

ii) **Log ratio:**To make the distribution more symmetric (normal-like) log-transformation can be applied. The most commonly used log-transformation is 2-based (*log* 2), but it does not matter whether you use natural logarithm (*loge*) or base 10 logarithm (*log*10) as long as it is applied consistently to all the samples. The formula for the calculation of *log*2-transformation is:

$$\text{Log ratio} = log_2(\text{Intensity ratio}) = log_2\left(\frac{Cy3'}{Cy5'}\right)$$

After the log-transformation, unchanged expression is zero, and both unregulated and down-regulated genes can take values from zero to infinity. Log ratio has some nice properties compared with other measures of expression. It makes skewed distributions more symmetrical, so that the picture of variation becomes more realistic. In addition, normalization procedures become additive. For example, we have the following intensity ratio results for two replicates (A and B) after normalization:

$$Gene\,A = \frac{120}{60} = 2.0$$

$$Gene\,B = \frac{30}{60} = 0.5$$

The mean of these replicates is 1.25 instead of 1, which would have been expected. If the 2-based logarithmic transformation is applied, the log ratios are:

$$Gene\,A = log_2\left(\frac{120}{60}\right) = 1.0$$

$$Gene\,B = log_2\left(\frac{30}{60}\right) = -1.0$$

The mean of these log ratios is 0, which corresponds to the mean intensity ratio of 1. Although log-transformation is not always the best choice for microarray data, it is used because the other transformations lack this handy additive property. One downside of the log-transformation is that it introduces systematic errors in the lower end of the expression value distribution.

3. **Replicates:** If the experiment includes replicates, their quality can be checked with simple methods, using scatter plots and pairwise correlations or hierarchical clustering techniques.When the distribution of the intensity values is skewed, the median characterizes the central tendency better than the mean. The median and mean can also be used to check the skew ness of the distribution. For symmetrical distributions mean and median are approximately equal.

4. **Outliers and Filtering bad & uninteresting data:** Outliers in chip experiments can occur at several levels. There can be entire chips, which deviate from all the other replicates. Or there can be an individual gene, which deviates from the other replicates of the same gene. Outliers should consist mainly of quantification errors. In practice, it is often not very easy to distinguish quantification errors from true data, especially if there are no replicate measurements. If the expression ratio is very low (quantification errors) or very high (spot intensity saturation), the result can be assumed to be an artifact, and should be removed. Most of the actual outliers should be removed at the filtering step (those that have too low intensity values). This is often equivalent to a filtering, where observations with too low or high intensity values are excluded from further analyses.

5. **Linearity:**Linearity means that in the scatter plot of channel 1 (red colour) versus channel 2 (green colour), the relationship between the channels is linear. It is often more informative to produce a scatter plot of the log-transformed intensities, because then the lowest intensities are better represented in the plot. In this kind of a plot, the data points fit a straight line, if the data is linear.Another way to test linearity is to plot the log ratio versus the intensity of one channel. This should be done for both channels independently.

STATISTICAL ANALYSIS

The basic concepts of clustering are to try to identify and group together similarly expressed genes and then try to correlate the observations to biology. The idea is that co-regulated and functionally related genes are grouped into clusters. Clustering provides the framework for this analysis.

Principles of clustering

1. Clustering organizes the data into a small number of homogeneous groups. The methods are used to find similar expression motifs irrespective of the expression level. Therefore, both low and high expression level genes can end up in the same cluster if the expression profiles are correlated by shape.

2. Clustering methods can be grouped as supervised and unsupervised. Supervised methods assign some predefined classes to a data set, whereas in unsupervised methods no prior assumptions are applied.

Different types of clustering involves Hierarchical clustering, K-means, self-organizing maps (SOMs), principal component analysis (PCA), and K-nearest neighbor (KNN) have been commonly used. There are also other methods, such as Correspondence Analysis, Fuzzy K-Means Clustering Multidimensional scaling (MDS), minimum description length (MDS), gene shaving (GS), decision trees, and support vector machines (SVMs).

1. Hierarchical clustering

The hierarchical clustering can be represented as a tree, or a dendrogram. Branch lengths represent the degree of similarity between the genes. Hierarchical clustering consists of two separate phases. The hierarchical clustering is performed as follows; Each gene is assigned to a cluster of its own. The closest pair of clusters is found and merged into a single cluster. The distances (similarities) between the new clusters are computed and each of the old clusters using either single, average or complete linkage method. Above steps are repeated until all genes are clustered.

In single linkage clustering, the distance between one cluster and another is considered to be equal to the shortest distance from any member of one cluster to any member of the other cluster. In complete linkage clustering, the distance between one cluster and another cluster is considered to be equal to the longest distance from any member of one cluster to any member of the other cluster. In average linkage, the distance between one cluster and another cluster is considered to be equal to the average distance from any member of one cluster to any member of the other cluster.

2. Self-organizing map

Kohonen's self-organizing map (SOM) is a neural net that uses unsupervised learning. SOM tries to learn to map similar input vectors (gene expression profiles) to similar regions of the output array of nodes. The method maps the multidimensional distances of the feature space to two-dimensional distances in the output map. In SOMs, the number of clusters has to be predetermined. The dimensions of the two-dimensional grid or array give the value. One has to experiment with the actual number of clusters and find such array dimensions that there is a minimum number of poorly fitting genes in the clusters.

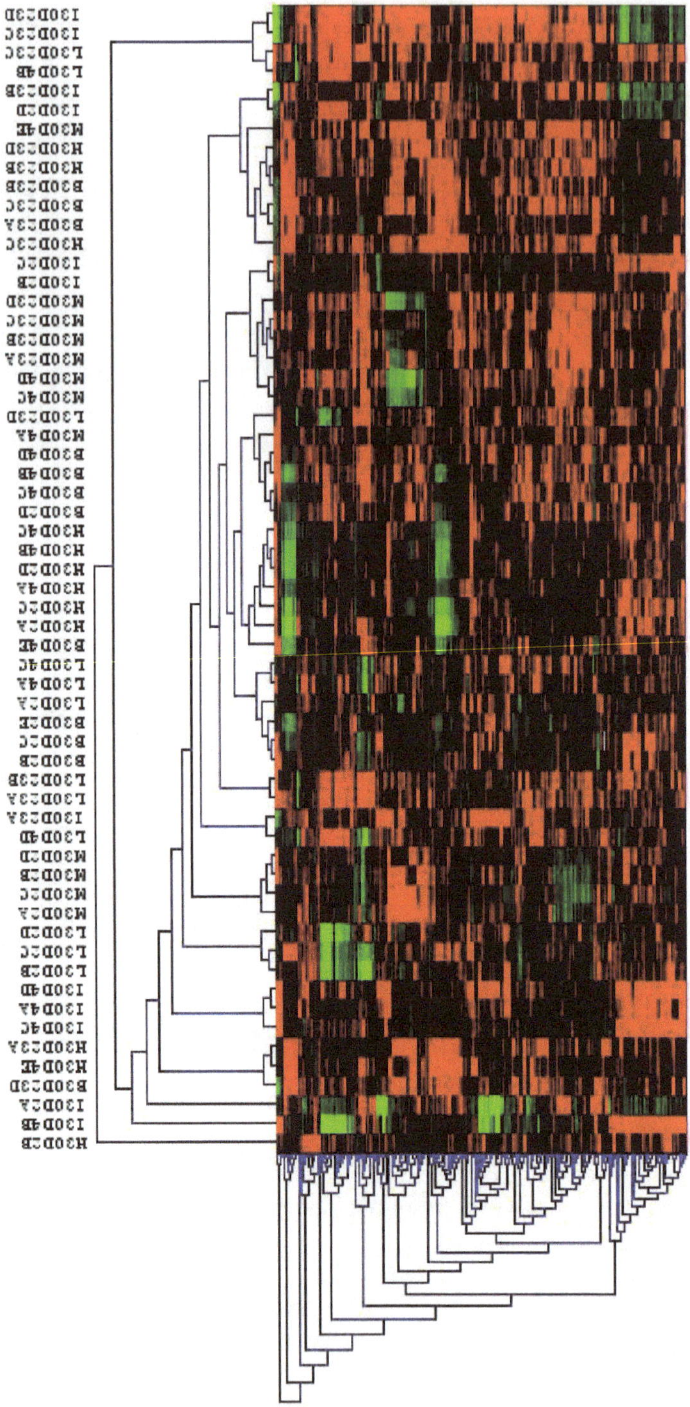

Fig. 4: An Example of Hierarchical clustering of *C. carpio* obtained using MeV software

The actual number of clusters is also difficult to assign, but the square root of the number of genes is a good initial estimate.

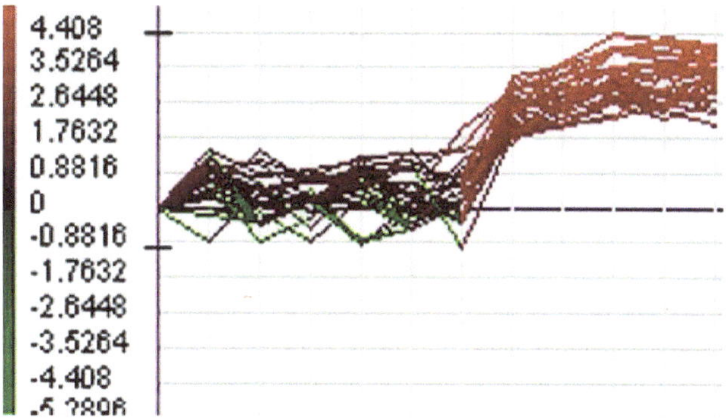

Fig. 5: An example of self-organizing map (*Source*; Hovatta *et al.*)

3. K-means clustering

K-means is a least-squares partitioning method for which the number of groups, K, has to be provided. The algorithm computes cluster centroids and assigns each object to the nearest centroid. However, it is also possible to estimate K from the data, taking the approach of a mixture density estimation problem. The genes are arbitrarily divided into K centroids. The reference vector, *i.e.* location of each centroid, is calculated. Each gene is examined and assigned to one of the clusters depending on the minimum distance. The centroid's position is recalculated. Above steps are repeated until all the genes are grouped into the final required number of clusters.

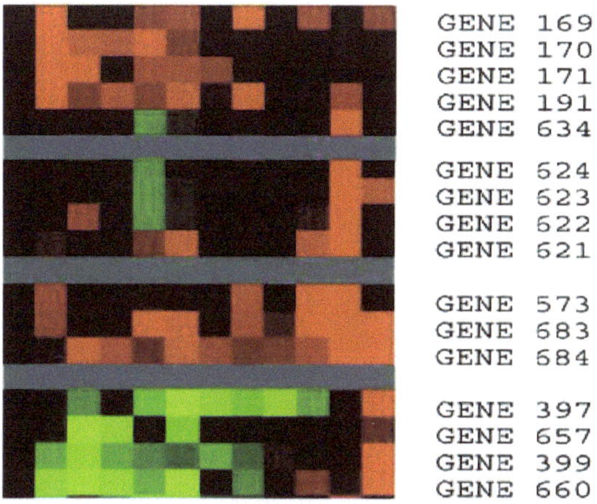

Fig. 6: An example of k-means clustering with four clusters (*Source*; Hovatta et. al.)

K-means partitioning is a so-called NP-hard problem, thus there is no guarantee that the absolute minimum of the objective function has been reached. Therefore, it is good practise to repeat the analysis several times using randomly selected initial group centroids, and check whether these analyses produce comparable results.

4. Principal component analysis

Objectives of principal component analysis (PCA) are to discover or to reduce the dimensionality of the data set and to identify new meaningful underlying variables. PCA transforms a number of (possibly) correlated variables into a (smaller) number of uncorrelated variables called principal components. The basic idea in PCA is to find the components that explain the maximum amount of variance possible by n linearly transformed components. The first principal component accounts for as much of the variability in the data as possible, and each succeeding component accounts for as much of the remaining variability as possible. PCA can be also applied when other information in addition to the actual expression levels is available (this applies to SOM and K-means methods as well).

5. Correspondence Analysis

Correspondence analysis is an explorative method to study associations between variables. It directly visualizes associations between genes and hybridizations.

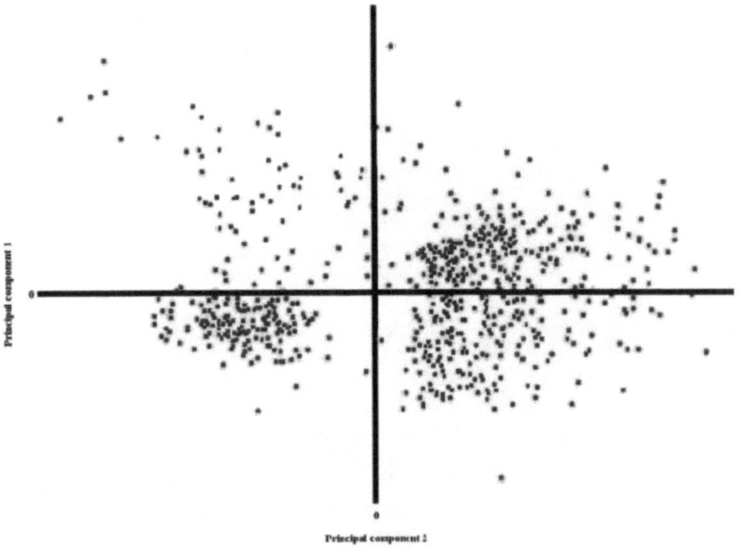

Fig. 7: An example of principal component analysis. The two most significant principal components have been selected as the axes of the plot (*Source*; Hovatta *et al.*)

Unlike many other methods, CA does not require any prior choice of parameters. Like principal components, it displays a low-dimensional projection of the data. However, in this case, both genes and samples can be projected onto the same space, revealing associations between them. Correspondence analysis requires an expression matrix with no missing values. Therefore, any missing values have to be imputed first. We use the k-nearest neighbor's algorithm to impute missing values. The only user input in the initialization dialog is the desired number of neighbors for imputation. Genes that lie close to one another on the plot tend to have similar profiles, regardless of their absolute value. The same is true for samples. If some genes and samples lie close to one another on the plot, then these genes are likely to have a high expression in the nearby samples relative to other samples that are far away on the plot. On the other hand, if a set of genes are on the opposite side of the plot from a set of samples relative to the origin, then the expression of that set of genes is likely to be depressed in those samples relative to samples that might be positioned close to those genes. The farther the points are from the origin, the stronger the association between genes and samples. Correspondence analysis works by decomposing a matrix of chi-squared values derived from the rows and columns of the expression matrix. The first two or three axes are the most informative in showing associations among genes and experiments. The amount of information explained by a given axis is quantified by its inertia, which may be thought of as the proportion of the total chi-squared value of the matrix explained by that axis. The inertia values are provided under the corresponding node under the main COA analysis node.

REFERENCES

Ames, B., and Hartmann, P. 1963. The histidine operon. *Cold Spring Harbor Symposium Quantitative Biology*, 28: 349-356.

Anderson, D.M., Kulis, D., Erdner, D., Ahn, S., Walt, D. 2006. Fibre optic microarrays for the detection and enumeration of harmful algal bloom species. *Afr. J. Mar. Sci.* 28(2): 231-235.

Bonnet E, Fostier A . and Bobe J. 2007. Microarray-based analysis of fish egg quality after natural or controlled ovulation. *BMC Genomics* 8(55):17.

Bobe, J., Montfort, J., Nguyen, T. and Fostier A. 2006. Identification of new participants in the rainbow trout (Oncorhynchus mykiss) oocyte maturation and ovulation processes using cDNA microarrays. *Reproductive Biology and Endocrinology*. (4):39, 16.

Clark, M.D., Hennig, S., Herwig, R., Clifton, S.W., Marra, M.A., Lehrach, H. and Johnson, S.L. 2001. An Oligonucleotide Fingerprint Normalized and Expressed Sequence Tag Characterized Zebrafish cDNA Library. *Genome Research*11(9):1594-1602.

Derome, N., Duchesne, P., Bernatchez, L. 2006. Parallelism in gene transcription among sympatric lake whitefish (Coregonus clupeaformis Mitchill) *Ecotypes Mol. Ecol.*, 15(5):1239-1249.

Greenacre, M.J. 1984. Theory and Applications of Correspondence Analysis. Academic Press, London.

Holm, L. 1986. Codon usage and gene expression. *Nucleic Acid Res.*, 14: 3075-3087.

Lafay, B., Lioyd, A.T., Mc Lean, M.J., Devine, K.M., Sharp, P.M. and Wolfe, K.H. 1999. Proteome composition and codon usage in spirochaetes: species-specific and DNA strand-specific mutataional biases. *Nucleic Acids Res.*, 27:1642-1649.

Larkin, Patrick, Villeneuve, Daniel L., Knoebl, Iris, Miracle, Ann L., Carter, Barbara J., Liu, Li, Denslow, Nancy D., Ankley and Gerald, T. 2007. Development and Validation of a 2,000-Gene Microarray for the Fathead Minnow (Pimephales Promelas). *Environ. Toxicol. Chem.* 26(7):1497-1506.

McLean, T.I. and Crawford, D.L. 2006. Towards a Genetic Understanding of Karenia brevis: Using Microarrays to Discover Gene Expression Patterns as They Relate to K. brevis Ecology EOS, *Transactions, American Geophysical Union*. 87(36): suppl.

Medigue, C., Rouxel, T., Vigier, P., Henaut, A. and Danchin, A.1991. Evidence for horizontal gene transfer in *Escherichia coli* speciation. *Journal of Molecular Biology*, 222: 851-856.

Moszer, I., Rocha, E.P.C. and Danchin, A. 1999. Codon usage and lateral gene transfer in *Bacillus subtilis* genome. *Microbiology*, 141: 261-268.

Mclnerney, J.O. 1998. Replicational and transcriptional selection on codon usage in *Borrelia burgdorferi*, *Proc. Natl. Acad. Sci.* USA, 95: 10698-10703.

Mclnerney, J.O. 1998. GCUA: General Codon Usage Analysis. *Bioinformatics*, 14:372-373.

Romero, H., Zavala, A. and Musto, H. 2000. Codon usage in Chlamydia trachomatis is the result of strand-specific mutational biases and a complex pattern of selective forces. *Nucleic Acid Res.*, 28: 1084-2090.

Roy, A.K. and Rath, S.N. 2007. Use of Correspondence Analysis in Gene Expression Study *of Labeo rohita*. 12th National Conference of Gwalior Academy of Mathematics and Sciences (GAMS) and All India Symposium on Computational Biology at Gwalior, (Abs. No. CB -20).

Roy, A.K., Martha, S.R. and Rath, S.N. 2007. Correspondence Analysis of Codon Usage in *Labeo rohita*. International Symposium on Chromosome to Genomes (c2g) at CCMB, Hyderabad, (Abs. No. 9)

Tsoi, S.C.M., Cale, J.M, Bird, I.M., Ewart, V., Brown, L.L., Douglas, S. 2003. Use of Human cDNA Microarrays for Identification of Differentially Expressed Genes in Atlantic Salmon Liver during Aeromonas salmonicida Infection. Marine Biotechnology [Mar. Biotechnol.]. 5(6):545-554.

Thioulouse, J., Chessel, D. and Dolédec, S. 1997. ADE-4: A multivariate analysis and graphical display software. *Stat. Comput.*, 7: 75-83.

Weidinger, G., Thorpe, C.J., Wuennenberg-Stapleton, K., Ngai, J., Moon R.T. 2005. The Sp1-Related Transcription Factors sp5 and sp5-like Act Downstream of Wnt/beta-Catenin Signaling in Mesoderm and Neuroectoderm Patterning. *Current Biology*. 15(6):489-500.

8

Significance of Insect Genomics in Agro Ecosystems

R.W. Alexander Jesudasan

A genome is an organism's complete genetic signature and the collection of information that an organism can pass on to its offspring. Genomics, the study of the structure, function, and evolution of entire genomes is principally dealing with genetic information which helps us to understand how organisms and ecosystems work, how theymalfunction, or how they might be melded to human needs. The model organisms used in genomics have, in general, small genomes and short generation times. To meaningfully analyze the gene-to-phenotype relationships, a large amount of sequencing information is necessary. Progress in DNA sequencing technologies leads to accumulate Information about structures of genomes (Mohan Kumar, 2007).

HISTORY OF INSECT GENOMICS

Commencing from the publication of the entire nucleotide sequence of the genome of *Haemophilus influenzae* in 1995, a complete genome sequence of many organisms, including lower and higher eukaryotes, have been reported. The cost-effective way to get sufficient sequence information to address the questions is to sequence a high-quality reference genome and then map sequences from other genomes onto the reference sequence. Insects constitute the vast majority of known species with their importance including biodiversity, agricultural, and human health concerns. Hence, a study of Insect genomics is imperative to elucidate information to other organisms.

INSECT GENOMICS

The information on structures, organization, and composition of genes are deposited in the insect genome databases which can be easily accessed through the web world. Information on the gene structures in the entire genome has accelerated the process of identification and characterization of genes responsible for various biological phenomena. Development of more comprehensive approaches for functional analysis of genes such as a large-scale gene disruption, transcriptome, and proteome analysis was possible through the understanding of sequences of entire genomes. These methods target a large number of genes at a time and demonstrate the useful combination of conventional methods with genome information. Consequently, a large quantity of knowledge on individual genes that are involved in a variety of biological processes has been accumulated and has facilitated understanding of the whole genetic systems in various organisms.

The sequencing of the *Anopheles gambiae* genome is a watershed event in genomics for two reasons. First this species is of sufficient phylogenetic distance from the previously sequenced *Drosophila melanogaster* to provide the best view to date of changes in genome organization and composition across the insects. The second reason that the *A. gambiae* genome is a landmark is that *Anopheles* is the first animal to be sequenced, other than ourselves, whose actions have a strong direct impact of human lives. In the near future such "applied" genomic projects will probably become the norm, as agencies involved with human health and agriculture develop plans to sequence key pests and beneficial species. This trend is particularly evident in insect genomics. The next two species in the insect genome study envisaged is the honey bee, *Apis mellifera* and silkworm moth, *Bombyx mori* because of their longstanding relevance in agriculture. Other insect candidates, include of that of another mosquito, *Aedes aegypti*, the medfly, *Ceratitis capitata*, and flour beetle, *Tribolium castaneum*, the polyphagous pests, *Helicoverpa armigera*, *Spodoptera litura*, the rice gall midge, *Orseolia oryzae*, the fruit fly, *Bactrocera* spp. and the polyphagous whitefly, *Bemisia tabaci*.

Whole-genome sequencing for various insect taxa was undertaken with a view to compare and contrast genome size, current genetic knowledge, species diversity, and the human impact of insects from 11 different insect orders. The outcome of the study indicated strengths in all criteria for the holometabolous insect orders viz., Hymenoptera, Diptera, Lepidoptera and Coleoptera. Further it was inferred that Coleoptera are the most speciose worldwide, but have slightly lower economic impact than Lepidoptera, Hymenoptera and Diptera. Beyond the holometabolous insects, the suborder Homoptera stands out for having species with generally small genomes and great economic and agricultural importance. The representatives from the primitive insect orders, Thysanura and Odonata, while valuable from the standpoint of phylogenetic breadth, have fared poorly compared to other orders.

The emphasis in these criteria on insects with recognized human impact and ecological importance is not meant to negate the value of model insect species as sequencing candidates. Model insect genomes can provide general insights into biological mechanisms, gene structure and function, and the conserved evolutionary process that select for certain genetic traits. Thus, model organisms, as illustrated by species of *Drosophila*, yield invaluable insights for all insect genomes.

Recognizing the success of the completed and ongoing dipteran genome projects, several working groups have formed to develop and promote genome projects in new insect groups. Within the Hymenoptera, an international genomics effort has been emerging for several years around the honey bee, arguably the best studied and economically most important member of this group. New applications of functional genomic techniques described foretell the many ways researchers will use genomic data to answer basic and applied questions in this species. As one example, a successful application of RNA interference methods in honey bee embryos and brains is worthy of mention.

Within the Lepidoptera, an international genomics effort has centered on the economically important silkworm moth. The International workshop of Lepidopteran genomics held at Tsukuba, Japan in 2002 focussed on key aspects of this genome project, most notably the integration of large-insert libraries, expressed sequence tags (ESTs), and applications of transgenic technologies.

Table 1: Insect orders evaluated for sequencing priority (After Evans & Gundersen-Rindal, 2003)

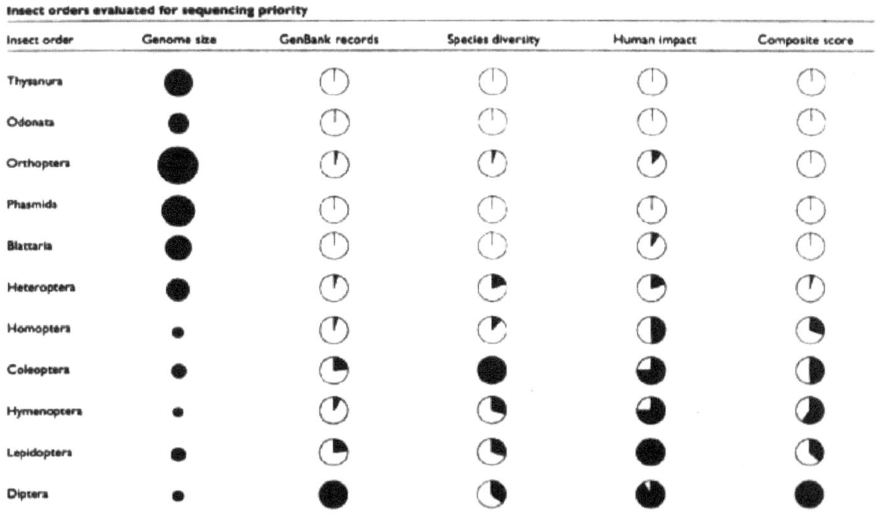

Insect orders are listed according to the phylogenetic tree shown in Figure 1 and show relative genome size (weighted mean, with size corresponding to the area of the circle). Also shown are the number of protein records in GenBank, number of worldwide species, and human impact estimate (see text for further details). In each case the proportion of the circle that is black indicates the number relative to the order with the highest value for each measure (the highest ranking order being shown with a filled circle). The last column shows a composite ranking of orders assuming that equal weight is given to each of these criteria.

The International Lepidopteran Genome Project has been charged with applying new technologies to compare the genomes of a growing list of agriculturally important moths and butterflies. Among these, the crop-feeding heliothine moths have long been appreciated as significant genome candidates.

Similar efforts in other economically important orders of insects such as Coleoptera and suborder Homoptera are gaining momentum. Well-defined and concerted research efforts, combined with advancing technologies and access to post-genomic tools and date, will accelerate the advances in these taxa (Table 1) (Evans & Gundersen-Rindal, 2003).

MOLECULAR STUDIES ON CROSS-TALK BETWEEN WOUND SIGNALLING THROUGH INSECT HERBIVORY AND OTHER PLANT DEFENCE RESPONSES

It has been shown that the initial wounding caused by insect feeding in plants finally leads to the production of proteinase inhibitors in the plant cell. The jasmonate signalling system mediates this process. The wounding causes the production of prosystemin which is a protein, which gets processed leading to the formation of systemin. The systemin molecule is systemic in nature, which means that it is readily transported from one part of the plant to the other through the conducting phloem tissue. The systemin molecule induces the synthesis of jasmonates in the cell by initiating a cascade of metabolic events. Jasmonates are able to induce a large number of genes in plants such as the proteinase inhibitor I and II genes and their expression is stimulated by wounding which may be due to simple mechanical shearing of leaves or herbivore-induced injury.

The methyl ester of jasmonic acid is highly volatile and it is readily transmitted through air to other plants growing in the neighbouring area, and induces insect resistance in them.

Further wound sites caused by insect herbivory are easily accessible to penetration by pathogens. A whole range of evidence points to the existence of cross-talk between wound- and pathogen-activated signalling pathways. Some of these regulatory interactions actually operate on the production of signals implication in different pathways (Fig. 1) (Swamy, 2000; Leon *et al.,* 2001).

AN ENEMY'S ENEMY IS A FRIEND - A MOLECULAR APPROACH

Apart from the direct way of defending themselves, plants are also able to defend themselves by indirect methods. Plants can respond to damage caused by herbivorous insects by producing volatile chemical compounds that attract natural enemies including the carnivorous insects and parasitoids thus facilitating the control of the insect pests.

By using genomics to unravel the signal responses and gene expression cascades involved in such response, it may be possible to enhance the endogenous defense mechanism of crops.

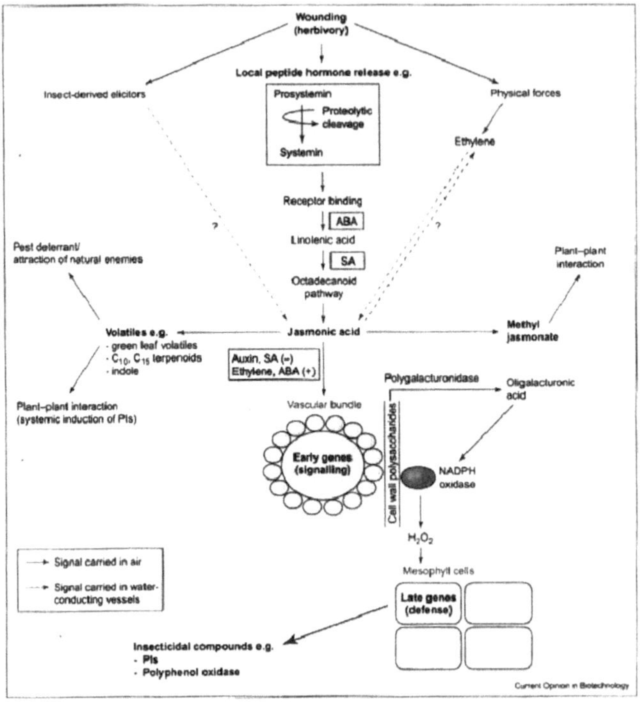

Fig. 1: Generalised overview of the plant wounding response and signalling molecules due to Insect herbivory (After Ferry *et al.*, 2004)

STATUS OF RICE GALL MIDGE BIOTYPES

The rice gall midge, *Orseolia oryzae* is an important insect pest of rice in South and Southeast Asia. Through the use of molecular tools biotypes of this insect has been reported and distinguished. In India, six distinct biotypes have been characterized, whereas in South China four biotypes have been recorded. A distinct link between the resistance gene in the host plant and biotypes exists. Based on reaction against resistant genes *Gm1* and *Gm2*, Indian biotypes designated as 1, 2, 3 and 4 show a similar pattern of reaction to that displayed by Chinese biotypes. Extensive cultivation of resistant varieties and mass migration appear to be the cause of biotype evolution (Fig. 2) (Pasalu *et al.,* 2004).

MOLECULAR IDENTIFICATION OF FRUIT FLIES, *BACTROCERA* SPP.

A species complex of *Bactrocera* has been reported to cause infestation to horticultural crops such as mango and guava. Since there has been problems pertaining to species identification using morphotaxonomy owing to the existence of polymorphism, sex and stage of development, molecular markers such as ITS 1, AFLP, mtCOI and nuclear gene period have been employed for this purpose (Ashokan *et al.,* 2007)

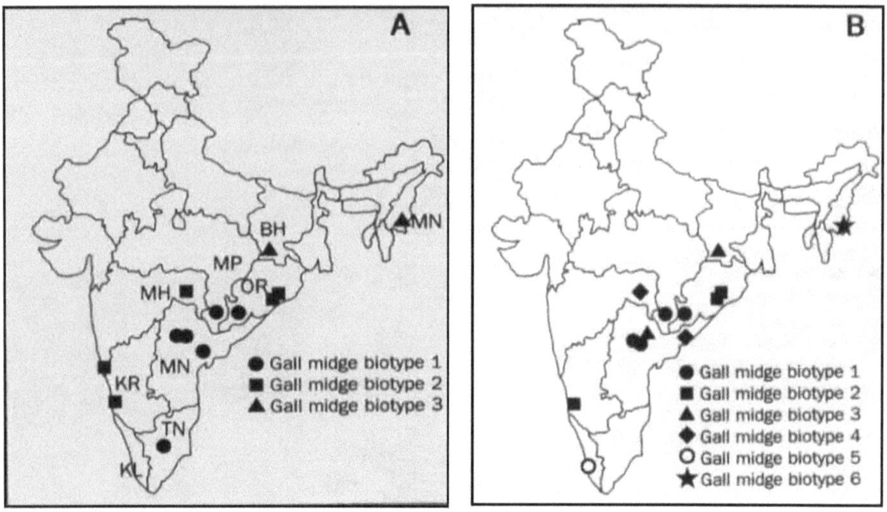

Fig. 2: Distribution of gall midge biotypes in India (A) during 1970-1983 & (B) during 1998 (After Pasalu *et al.,*, 2004)

MOLECULAR CHARACTERIZATION OF THE WHITEFLY, *BEMISIA TABACI*

'The Christmas invasion' about whitefly havoc is interesting and alarming not only for the Americans, but of global concern causing severe damage to a variety of crops annually in different niches. The morphological character and behaviour of the pest is so tuned genetically with its evolutionary stage that it easily protects its body from pesticide droplets (Singh, 2007). Further, the pest status of the cosmopolitan whitefly, *Bemisia tabaci* is not only recognized as a vector of several viral diseases but also due to its impact owing to honeydew exudation by nymphs and adults resulting in the fouling of the lint in cotton making it difficult to process resulting in substantial loss to cotton industry (De Barro, 1995).

The existence of polymorphism or intraspecific diversity again in whiteflies has intensified the problem owing to wrong identification of the taxa. The advent of molecular taxonomy and bioinformatics has resulted in the detection of several biotypes in *B. tabaci* and through this effort it has since been possible to understand the features of each of them in order to apply them in the insect pest management programmes.

CONCLUSION

In spite of the importance of insects, our knowledge of their biology is still far from complete. At present, however, there are highly exciting developments occurring within the field of insect research, because the genomes from 24 insect species have recently been sequenced, or are in the process of being sequenced.

The outbreak of genome sequences is also likely to impact genetic studies more broadly. Researchers studying insect genomics can look forward to using these shared traits to better address general problems in medicine, biotechnology, agriculture, and evolutionary biology. All these developments will certainly lead to a significant reduction in worldwide pesticide use, which will be beneficial for human health and the environment. This means that insect pest control soon will enter the genomic era with all its surprises and discoveries.

REFERENCES

Ashokan, R., Krishna Kumar, N. K. and Abraham Verghese. 2007. Molecular identification of fruit flies, *Bactrocera* spp. (Diptera: Tephritidae) using mitochondrial cytochrome oxidaseI.*Current Science,* 93 (12): 1668-1669 .

Evans, J.D. and Gundersen-Rindal. 2003. Beenomes to Bombyx : future directions in applied insect genomics. *Genome Biology,* 4: 107.

Ferry, Natalie, Martin, G. Edwards, John, A. Gatehouse and Angharad, Gatehouse, M.R. 2004. Plant-insect interactions: molecular approaches to insect resistance. *Current opinion in Biotechnology*, 15:1-7.

Pasalu, I.C., Bing-Chao Huang, Yang Zang and Yu-Juan Tan. 2004. Current Status of rice gall midge biotypes in India and China. In New Approaches to gall midge resistance in rice. *Eds. J. Bennett et al.,Proceedings of the International Workshop*, 131-138.

9

Bioinformatics Tools in Fish Molecular Modeling

Grinson George, K.Sarma, S. Dam Roy and Mukunda Goswamy

Bioinformatics can be defined as the application of computer technology for the management and analysis of biological data using computer to gather, store, analyze and merge the biological data. It consists of the organization of many kinds of large-scale biological data, particularly that based on DNA and protein sequence, into web-accessible databases and the methods required analyzing these data. Bioinformatics is a fast growing field within the biological sciences that was developed because of the need to handle large amounts of genetic and biochemical data. The goal of bioinformatics is to uncover the wealth of biological information hidden in the mass of data and to obtain a clearer insight into the fundamental biology of organism. This knowledge can have profound impact on various fields such as human health, agriculture, environment, energy and biotechnology. Bioinformatics involves use of various biological databases, which are archives of consistent data that are stored in a uniform and efficient manner. Biological databases are large collections of data that are relatively different to maintain outside the centers and institutions that produce them. These data are traditionally accessed and analysed using browser-based World Wide Web (www) interfaces. There are two types of databases i.e. Primary database and Secondary database. Primary databases contain information and annotations of DNA and Protein sequences, DNA and Protein structure and DNA and Protein expression profiles catalogued for the end users. Secondary databases contain results of analysis on primary resources including information on sequence patterns or motifs, variants, mutations and evolutionary relationships for further use by an interested researcher.

Web browsers provide easier integration and interoperability between bioinformatics applications and the data they require in a faster mode. But problem occurs when large amounts of data need to be retrieved and analyzed for a specific purpose. This often proves to be a tedious and impractical job. Moreover, just retrieving or analyzing a particular nucleotide or protein sequence rarely completes the research as intended. When performing the operations mentioned above using a web browser, researchers are forced to repeat the troublesome tasks of searching and copying the results for further analysis for the specific problem they are addressing to. The problem gets solved if database information retrieval and analysis services get linked in a compatible mode, so that, search results from one database can be used as the base of a search another and the results of which are then analyzed.

Thus a bioinformatics work environment has been created by downloading and installing the necessary database content and services (such as retrieval and analysis programme) - working locally using information available globally. This has the advantage of data processing otherwise requiring huge amount of manual operations. However, the amount of disk space required to store biological sequence database can be huge, often exceeding several terabytes, requiring several hours, if not days, to complete the analysis as intended by a researcher on a specific problem.

DATA AS REFERRED UNDER THE DOMAIN OF BIOINFORMATICS

A large variety of data have now been incorporated into biological database and can be considered to be part of the domain of bioinformatics: DNA sequence, RNA sequence, Protein sequence, Expressed sequence tags (ESTs), Structural data from X-ray crystallography and nuclear magnetic resonance (NMR), Data from the analysis of transcription (transcriptomics), Data from proteomics experiments including metabolic, phosphor-proteomics, two hybrid analysis and the like.

BIOINFORMATICS- APPLICATION MODE

Bioinformatics has wide applications in fish molecular modelling. It is mainly applied for manipulation and analysis of individual nucleic acid and protein sequences as intended by a researcher. The maintenance of sequence data and other above mentioned data types in database are designed to allow their easy manipulation and inspection, often in the context of access on the Internet at a faster operational mode. The best known sequence database include EMBL and Genbank, which contain all known sequence data and Swiss-Prot and TrEMBL (now combined into the overarching UniProt), which contain protein sequences, including those derived from translation of putative protein-coding genes in DNA sequence used for molecular modeling works. It is mostly used for analytical methods, particularly for sequence similarly searching, sequence alignment and to identify unknown sequences and assign functions, identification structural or functional motifs, protein domains; thus having a great role in molecular modeling and related aspects. It has also applications in establishing beyond doubt phylogenetic relationships to identify evolutionary relationships by multiple alignments, mostly of protein and DNA sequences. Bioinformatics has been used effectively for structural analysis (structural bioinformatics) such as prediction of secondary structure from protein sequence, prediction of protein tertiary structures from the known structures of homologous proteins (homology modeling) and others. Statistical analysis of transcription or protein expression patterns is worked using bioinformatics. The identification of clusters genes, whose transcription varies in a similar way in response to a particular disease state, is an important step in analyzing large amount of transcriptomic or proteomic data. The important role played by bioinformatics in development of large database systems (Entrez, Site Replication Service; SRS) to encompass many types biological information, allowing cross-comparisons to be made suiting the modern research avenues.

BIOLOGICAL DATABASES

Biological databases are the archives of consistent data that are stored in a uniform and efficient manner to be utilized by a researcher. These databases contain data from a broad spectrum of molecular biology areas. The databases dealt with are as follows:

i) Taxonomic

ii) Nucleotide and

iii) Protein Databases

Taxonomic Databases- An Over View

Taxonomic databases are Taxonomic Browsers maintained by National Centre for Biotechnology Information (NCBI) for the purpose of identification, classification and characterization of organisms. This database is hierarchical and sequence based, aiming to centralize the classification of all organisms represented in the database with at least one nucleotide or protein sequence from the concerned organism. Fish Base, online taxonomic database (www.fishbase.org) has information on nearly all of the fish species in the world – over 28,600, or 96 per cent of the total known to us maintained by World Fish Centre. The electronic encyclopedia provides essential information on everything from their biology, life history, ecology and disease to taxonomy, aquaculture and associated ecosystems. It also offers historical data reaching back 250 years, including the largest collection of fish museum and survey data on the Internet in digital form (over 1.9 million records). Similarly there is a large volume of information about corals in reef base available at ICLARM site.

Nucleotide Databases

Nucleotide databases are maintained by International Nucleotide Sequence Database Collaboration (INSDC) which is a joint operation by EMBL-Bank at European Bioinformatics Institute (EBI), DDBI (DNA Data Bank of Japan) at National Institute of Genetics and GenBank at NCBI for use by researcher's world wide. The nucleotide sequence databases are data resources accepting nucleic acid sequence data from the scientific community and making it fully available for the end users. The databases record every publicly known nucleic acid sequence carried out by researchers.

EMBL-BANK

i) Website address:- www.ebi.ac.uk/

ii) EMBL is the acronym for European Molecular Biology Laboratory.

iii) European Bioinformatics Institute (EBI) has created EMBL database for the end users.

iv) It constitutes of Europe's primary nucleotide sequence resources.

v) It works in international collaboration with GenBank (USA) and DNA Databank of Japan (DDBI), thus forming a very strong network.

vi) It contains direct submissions from individual researchers, genome sequence projects and patent applications for endusers.

EMBL contains various parameters like Databases, Tools, EBI groups etc. The databases include database browsing, literature, nucleotide, protein, structure, proteomic etc. and the tools include protein functional analysis, proteomic services, sequence analysis, similarly & homology, structural analysis and the like. The various operations are performed by using these tools in relative forms.

DDBJ

i) The network Began DNA data bank activities in 1986 at National Institute of Genetics.

ii) Functioning as international nucleotide sequence database in collaboration with EBI/EMBL & NCBI/GenBank, thus forming another crucial network.

iii) DNA bank of Japan is officially certified to collect DNA sequence from researchers & to issue internationally recognized accession number to data submitters- atypical cataloguing system.

iv) Exchange the collected data with EMBL/EBI & GenBank/NCBI on a daily basis for end users.

v) Search is done by FASTA and Blast for faster search.

vi) Analysis of data is done by ClustalW.

GENBANK

i) GenBank is a National Institute of Health (NIH) genetic sequence database.

ii) It is an annotated collection of all publicly available DNA sequences for end users.

iii) It has been created by National Centre for Biotechnology Information (NCBI).

iv) It maintains following databases and tools:

   a. Literature databases eg. PUBMED, OMIM,

   b. Entrez Databases eg. Conserved Domain Databases, 3D Domains.

   c. Nucleotide Databases eg. Expressed Sequence Tags (EST).

   d. Tools for Sequence analysis eg. BLAST.

   e. Tools for 3-D structure display and Similarity searching eg. Cn3D.

SEARCHING THROUGH NCBI

Type the query into the dialogue box and press 'GO'. Search can be done through many techniques like 'All database', 'website', Literature' etc.

ENTREZ DATABASES

CDD: The Conserved Domain Database (CDD) contains protein domain models imported from outside sources, such as Pfam and SMART, and curates at NCBI. CDD contains over 12,000 such models and is linked to other NCBI databases, including protein sequences, bibliographic citations, and taxonomy. Search can be done by selecting 'Conserved Domains' and typing the query eg. Retinol in the text dialogue box and clicking 'Go'.

MeSH: MeSH is the U.S. National Library of Medicine's controlled vocabulary used for indexing articles for MEDLINE/PubMed. MeSH terminology provides a consistent way to retrieve information that may use different terminology for the same concepts.

OMIM: OMIM stands for Online Mendelian Inheritence in Man. The database contains textual information and references. It also contains copious links to MEDLINE and sequence records in the Entrez system, and links to additional related resources at NCBI and elsewhere.

NUCLEOTIDE DATABASES

EST: EST is a division of GenBank that contains sequence data and other information on "single-pass" cDNA sequences, or Expressed Sequence Tags, from a number of organisms.

GSS: GSS stands for Genome Survey Sequence. It contains the following types of data.

i) Random "single pass read" genome survey sequences.

ii) Cosmid /BAC/YAC end sequences.

iii) Exon trapped genomic sequences.

iv) Alu PCR sequences.

v) Transposon –tageed sequences.

STS: STS stands for Sequence Tagged Sites. It is an NCBI resource that contains sequence and mapping data on short genomic landmark sequences or Sequence Tagged Sites.

TOOLS FOR SEQUENCE ANALYSIS

BLAST (Basic Local Alignment Search Tool)

It performs "local" alignment. It finds regions of local similarly between sequences. The program compares nucleotide or protein sequences to sequence databases and calculates the statistical significance of matches.

For performance BLASR, BLAST option is selected from NCBI home page.

Blast search box accepts a no. of different inputs and automatically determines the formal.

FASTA that is a sequence that begins with a single line description, followed by lines of sequencing data. The description line differs by a greater than sign, '>' in the first column.

The other identifiers are Accession no., Accession version or gi number eg. (p01013, AAA68881.1, 129295).

TYPES OF BLAST

i) Blastn: searches a nucleotide database using a nucleotide query.

ii) Blastp: searches a protein database using a protein query.

iii) Blastx: searches a protein database using a translated nucleotide query.

iv) Tblastx:searches a translated nucleotide database using a translated nucleotide query.

v) Tblastn: searches a translated nucleotide database using a protein query.

FASTA

Fasta is a Protein similarity search. It provides sequence similarity searching of query sequence against nucleotide & protein databases using fasta programs. It can be used for fast protein comparison or fast nucleotide comparison. This format contains a one line header followed by lines of sequence data. The sequences in fasta formatted files are preceded by a line starting with a ">" symbol. The first word on this line is the name of the sequence. The rest of the line is a description of the sequence. The remaining lines contain the sequence itself. Fasta files containing multiple sequences are just the same, with one sequence listed right after another. This format is accepted for many multiple sequence alignment programs.

COMMONLY USED SOFTWARE USED IN FISH MOLECULAR BIOLOGY

The selection of software depends upon the type of data to be generated. Not all the software can analyse all sorts of data. Some software is supplementary to some other software. The most commonly used software are PHYLIP, ARLEQUIN, MEGA, CLUSTAL W, CLUSTAL X, GENEDOC, TFPGA, DnaSP, NETWORK, DNASTAR, GENETOOL, BATWING, SPSS, StatXact, SEQUIN etc.

DATA TYPES

Data types can be of different types depending upon the nature of work which comprised of following types.

- DNA sequence
- Micrsaatellites
- Binary Data
- RELP
- RAPD
- Protein Sequence
- SNP's (Single Nucleotide Polymorphisms)
- UEP's (Unique Event Polymorphisms)

TYPES OF STRUCTURE

CLUSTALW

Clustalw is a multiple sequence alignment program for protein or nucleotides which is available at <ebi.ac.uk/clustalw> Here the multiple sequence alignment means an extension of pairwise alignment to incorporate more than two sequences at a time. It is often used in identifying conserved sequence regions across a group of sequences hypothesized to be evolutionarily related.

- It calculates the best match for the selected sequences so that differences, similarities and identities can be seen.
- Evolutionary relationship can be seen using cladograms and phylograms. Phylogram is a branching diagram (tree) assumed to be an estimate of a phylogeny; branch lengths are proportional to the amount of inferred evolutionary change. A Cladogram is a branching diagram (tree) assumed to be an estimate of a phylogeny where the branches are of equal length, thus cladograms show common ancestry, but do not indicate the amount of evolutionary "time" separating taxa.
- It can align either nucleotide or protein sequences. In the case of nucleotide sequences, it will align them as they are input – the program does not provide the option of specifying DNA strands.
- The program accepts sequences in the formats like-NBRF/PBR, EMBL/UniProt, Pearson (Fasta), GDE, ALN/ClustalW. The sequences can either be pasted into the web form or uploaded to the web form in a file.

CLUSTALX

Clustal X is a new windows interface for the ClustalW multiple sequence alignment program. It provides an integrated environment for performing multiple sequence and profile alignments and analyzing the results. The sequence alignment is displayed in a window on the screen. A versatile coloring scheme has been incorporated allowing you to highlight conserved features in the alignment. This is used for aligning the DNA sequences, which is necessary for further analysis in PHYLIP, ARLEQUIN and much other software. Clustal X (Thompson *et al.* 1997) is a version of Clustal W with a graphical user interface. The program is designed to (I) perform multiple alignments, (2) view the results of the alignment is facilitated by options that are not available in Clustal W (e.g., selecting a section of the alignment to be realigned with different gap penalties, while keeping the rest of the alignment fixed). The method Clustal uses to construct the alignment is called pairwise progressive sequence alignment. This heuristic method first does pairwise sequences alignment for all the sequence pairs that can be constructed from the sequence set. A dendrogram (guide tree) of the sequences is then done according to the pairwise similarly of the sequences. Finally a multiple sequence alignment is constructed by alignment sequences in the order, defined by the guide tree. The pull-down menus at the top of the window allow you to select all the options required for traditional multiple sequence and profile alignment. Alignment quality analysis can be performed and low-scoring segments or exceptional residues can be highlighted.

GENEDOC

GeneDoc is a full featured multiple sequence alignment editor, analyser and shading utility for windows. This is used for aligning the DNA sequences, which is necessary for further analysis in ARLEQUIN and much other software.

NETWORK

Network generates evolutionary trees and networks from genetic, linguistic, and other data. Network can then provide age estimates for any ancestor in the tree. This is used for forming the network of different haplotypes present in the population / populations and shows the mutation by which the haplotypes differ. This software also helps us find the TMRCA (Time to most recent common ancestor) when we feed the mutation rate. All the above mentioned data types can be handled by this software. The disadvantage of this software is that it gives a complex tree when sample size is more.

The PHYLogeny Inference Package (PHYLIP) is a package of programs for inferring phylogenies (evolutionary trees). It is available free over the Internet, and written to work on a many different kinds of computer systems as possible. The types of data which can be handled by PHYLIP are DNA sequences, gene frequencies, restriction sites (RFLP), distance matrices and binary data. This is mostly used for estimating genetic distance, consensus, neighbor joining, maximum likelihood (counts the number of specific pattern are grouped into a cluster) and maximum parsimony (the amount of heterogeneity is computed using the transition and transversion ratio and then those two sequences with the least heterogeneity is regarded as a group). The main disadvantage of this software is complex tree when sample size is more. This software can't handle microsatellite data and can't give phylogeny tree at population level. Methods that are available in the package include parsimony, distance matrix, and likelihood methods, including boot strapping and molecular consensus trees. Data types that can be handled include sequences, gene frequencies, restriction sites and fragments, distance matrices, and discrete characters.

ARLEQUIN

Arlequin is an exploratory population genetics software environment able to handle large samples of molecular data (RELPs, DNA sequences, microsatellites), while retaining the capacity of analyzing conventional genetic data (standard multi-locus data or mere allele frequency data). This software performs analysis at population level other than data analysis. This program incorporates a large set of methods and statistical tests in order to extract information on genetic and demographic features of collection of population samples. All the above mentioned data types can be handled by this software. This software does analysis at inter and intra-population level. The disadvantage of the software is it can't build up maximum likelihood and maximum parsimony methods. A variety of population genetics methods have been implemented either at the intra-population or at the inter-population level, and they can be conveniently selected and parameterized through a graphical interface. Arlequin has no equivalent in his field and will be extremely useful to analyze large data sets which are now available by the use of the latest molecular engineering techniques.

Arlequin has been designed to be able to handle different types of molecular or conventional (non-molecular or frequency-type) data it can also handle data either presented in the form of genotype frequencies, or the form of haplotype frequencies, as well as the possibility of treating codominant or recessive data (with the definition of a single recessive allele per locus). Molecular data can be analyzed by entering their definition (as DNA sequences, RFLP haplotypes, microsatellite profiles, or multilocus haplotypes), or by entering a distance matrix defining the relationships among the haplotypes (as was done with our previous AMOVA software). The data format is specified in an input file.

SPSS AND STATXact

These softwares are used for drawing multidimensional plots form gene, principal component analysis from genetic distances. It is also used for getting allele frequencies from microsatellite data. BATWING (Bayesian Analysis of Trees with Internal Node Generation).

This software is used for the calculation of TMRCA. This software can handle all above mentioned data types except DNA sequence. The software is being updated for taking DNA sequence.

TFPGA

Tools for Population Genetic Analyses: A windows program for the analysis of allozyme and molecular population genetic data. This program calculates descriptive statistics, genetic distances, and F-statistics. It also performs tests for Hardy-Weinberg equilibrium, exact tests for genetic differentiation, Mantel tests, and UPGMA cluster analyses. Additional features include the ability to analyze hierarchical data sets as well as data from either codominant markers such as allozymes or dominant markers such as AFLPs or RAPDs.

TREE VIEW

Tree-View provides a simple way to view the contents of a NEXUS, PHYLIP Hennig86, Clustal, or other format tree file. While PAUP and MacClade have excellent tree printing facilities, there may be times you just want to view the trees without having to load the data set they were generated from. The PHYLIP package contains tree drawing programs which offer a greater variety of trees than Tree View, but are somewhat clumsy to use.

BIOEDIT

It is a biological sequence alignment editor written for Windows 95/98/NT/2000/XP. An intuitive multiple document interface with convenient features which makes alignment and manipulation of sequences relatively easy on your desktop computer. Several sequence manipulation and analysis options and links to external analysis programs facilitate a working environment which allows you to view and manipulate sequences with simple point and click operations.

SEQUIN

Sequin is a stand- alone software tool developed by the NCBI for submitting and updating entries to the GenBank, EMBL, or DDBJ sequence databases. It is capable of handling simple submissions that contain a single short mRNA sequence, and complex submissions containing long sequences, multiple annotations, segmented sets of DNA, or phylogenetic and population studies.

DNASTAR IN MITOCHONDRIAL DNA SEQUENCE PROCESSING AND ALIGNMENT

An alignment programme seeks an arrangement that maximizes the net score by penalizing the alignment score for each gap and for each additional residue in a gap. For nucleic acid alignments, matching residues usually get a score of 1 and mismatches get a score of 0. The ultimate achievement would be to create an alignment in which, all the gaps represent the real deletion or insertion events that occurred during the divergence from a common ancestral sequence. We should attempt to minimize the number and size of the gaps while maximizing the extent of conserved blocks. A 'conserved block' is a region in which similar or identical nucleotides occur across all or most of the sequences.

Aligning a pair of sequences is not a computationally difficult process, and a variety of programmes exist to align sequence pairs. Multiple alignments are considerable more complex, and ClustalW or its updated version ClustalX is one of the best tools for creating multiple alignments. Clustal, like any other computer programme requires that the data it manipulates (the input file) must be in a format that it can recognize. The FASTA format can be recognized because the first line begins with the ">" character. This is followed by a single word (with 10 or few characters containing only letters and numbers) that Clustal will use as the name of the sequence in the multiple alignment that it creates. The steps involved in aligning the sequences using the software "DNASTAR" that works based on ClustalW are described below. DNASTAR was founded by University of Wisconsin Professor of Genetics Fredrick Blattner and his student colleague John Schroeder in the early 1980s. DNASTAR was among the early pioneers in bioinformatics. Lasergene, is the premier suite of software tools in DNASTAR for sequence assembly, editing, alignment, analysis and primer designing. The DNASTAR consists of pragrammes like MegAlign, EditSeq, GeneQuest, MapDraw, Primer Select, Protein and SeqMan.

SOFTWARE FOR PRIMER DESIGNING

Primer 3, CODEHOPE, Webprimers and genetool are mostly use d for PCR primer desgning.

PRIMER3

Primer3 is a widely used program for designing PCR primers (PCR = "Polymerase Chain Reaction"). PCR is an essential and ubiquitous tool in genetics and molecular biology. Primer3 can also design hybridization probes and sequencing primers steps involved in genetool are given at the end.

MEGA (MOLECULAR EVOLUTIONARY GENETICS ANALYSIS)

MEGA is an integrated tool for automatic and manual sequence alignment, inferring phylogenetic trees, mining web-based databases, estimating rates of molecular evolution, and testing evolutionary hypotheses. MEGA has become essential for reconstructing the evolutionary histories of species and multigene families, estimating the rates of molecular evolution, and inferring the nature and extent of selective forces shaping the evolution of genes and genomes. This need is now well recognized and has led to a greatly expanded scope for application of computational and statistical methods.

DNASP

DnaSP, DNA Sequence Polymorphism is a software package for the analysis of nucleotide polymorphism from aligned DNA sequence data. SnaSp can estimate several measures of DNA sequence variation within and between populations (in non-coding, synonymous or non-synonymous sites, or in various sorts of codon positions), as well as linkage disequilibrium, recombination, gene flow and gene conversion parameters. DnaSP can also carry out several tests of neutrality: Hudson, Kreitman and Aguade (1987), Tajima (1989), McDonald and Kreitman (1991), Fu and Li (1993), and Fu (1997) tests. Additionally, DnaSP can estimate the confidence intervals of some test-statistics by the coalescent. The results of the analyses are displayed on tabular and graphic form.

PROTEIN DATABASES

Protein databases are the most comprehensive source of information on protein. It is necessary to distinguish between universal databases covering protein from all species and specialized data collections storing information about specific families or protein of a group or an organism.

TYPES

i) Primary protein Databases like UniProt/Swiss-Prot.

ii) Secondary Protein Databases like Interpro.

iii) Specialised Protein Databases like GOA, ENZYME.

iv) Structure Databases like PDB.

Important software for proteomics. CN3D, SPDV, RASMOL, PSI-_ PREP, JPRED, PDB, SCOP, CATH, TrEMBL.

UNIPROT: It stands for Universal Protein resources. It is a central repository of protein sequences and function, created by joining the information contained in UniProt/ SwissProt.

PDB: PDB stands for protein Data Bank. The RCSB PDB provides a variety of tools and resources for studying the structures of biological macromolecules and their relationships to sequence, function, and disease.

PDB VIEWER: It provides an introduction to macromolecular modeling with Deep View, including review of many basic concepts in protein structure. Structure files are used for viewing, and then carry out exercises in manipulating, analyzing and comparing protein structures.

Different Molecular Viewers are as follows:

i) King http://kinemage.biochem.duke.edu/software/king.ph

ii) Jmolhttp://jmol.sourceforge.net/

iii) WebMolhttp://www.cmpharm.ucsf.edn/_walther/webmol.html

iv) QuickPDB<http.//www.rcsb.org/robohelp_f/viewers/thequickpdb.html>

v) ProteinWorkshop http://www.rcsb.org/robohelp/viewers/proteinworkshop.htm

Other options for plugin viewers are:

i) RasMol http://tsmb06.cryst.bbk.ac.uk/tech/rasmol/index.html#chime

ii) Swiss PDB Viewer http://www.usm.maine.edu/-rhodes/SPVTut/

SWISS PDB VIEWER: Deep View (formerly called Swiss-Pdb Viewer) is a friendly but powerful molecular graphics program. It is designed for use with computing tools available from the Expert Protein Analysis System or ExPASy Molecular Biology Server in Geneva, Switzerland.

i) It allows us to build models, by giving an amino-acid sequence.

ii) It can find hydrogen bonds within proteins and between proteins and ligands.

iii) It allows us to view several proteins simultaneously and superimpose them to compare their structures and sequences.

iv) It computers electrostatic potentials and molecular surfaces, and carries out energy minimization.

v) Getting started:- website www.rcsb.org

vi) Enter the name of protein name or PDB ID (contng. Four characters) into the search box. Then download Pdb file.

RASMOL: RasMol is a computer program written for molecular graphics visualization intended and used primarily for the depiction and exploration of biological macromolecule structures, such as those found in the Protein Data Bank. It is available for Windows, Macintosh and UNIX platforms.

Ca3D: Cn3D is a helper application that allows us to view 3-dimensional structures from NCBI's Entrez retrieval service. Ca3D runs on Windows, Macintosh, and Unix, Cn3D is a visualization tool for biomolecular structures, sequences and sequence alignments and has powerful annotation and alignment editing features. Cn3D displays structure-structure alignments along with their structure-based sequence alignments, to emphasize what regions of a group of related proteins are most conserved in structure and sequence. While working on it two windows appear: the main Cn3D structure window where the protein is displayed, and a sequence window that shows the protein chain's amino acid sequence.

When a single structure is loaded into Ca3D, the sequence viewer shows the sequences of all protein and nucleic acid chains in the structure. The color of each residue is coordinated between the structure and sequence windows: each letter of the sequence represents a residue in the structure, and always adopts the color of the backbone's alpha carbon (or phosphorus, for nucleotides), even if side chains are colored differently from backbone in the structure window Cn3D's sequence window also functions as an alignment viewer when displaying more than one structure or a structure to which multiple sequence have been aligned.

Software's Available at the following web sites:

Arlequin: http://antro.unige.ch/arlequin

Phylip: http://evolution.genetics.washington.edu/phylip

ClustalW: www.ebi.ae.uk/clustalw/

ClustalX: ftp://ftp-igbmc.u-strasbg.fr/pub/ClustalX/

MEGA: http://www.megasoftware.net

DnaSP: http://www.ub.es/dnasp/

TEPGA: http://www.marksgeneticsoftware.net/tfpga.htm

NETWORKING: http://www.flusus-engineering.com/sharenet.htm

Batwing: http://www.maths.abdn.ac.uk/ijw

Network: http://www.fluxus-engineering.com

SPSS: http://www.systat.com

Statxact: http://www.cyteL.com/new.pages/SX.2

BIOEDIT: http://www.mbio.nesu.edu/BioEdit/bioedit.html

Sequin: http://www.nebi.nlm.nib.gov.ilsprod.lib.neu.edu/Structure/CN3D/cn3D.shtml

RASMOL. www.umass.edu/microbio/rasmol/

SPDV: www.swissmodeLexpasy.org/spdbv/text/download.gtm

CATH: www.eathdb.info/

10

Structural Bioinformatics Goals and Challenges

Mayank Pokhriyal and M. Balakrishnan

Bioinformatics can be defined as "the collection, archiving, organization and interpretation of biological data" (Orengo *et al.,* 2003). By extension, structural bioinformatics deals with the collection, archiving, organization and interpretation of 3-D structural data of bio macromolecules. Structural bioinformatics emerged with the first crystal structures of haemoglobin and myoglobin. These structures were analyzed and contributed significantly to our understanding of the principles of life at a molecular level. A repository of all publicly accessible protein crystal structures, the Protein Data Bank (PDB) was established by Walter Hamilton *et al.,* at Brookhaven National Laboratory in 1971. Until the late 1980s, structural biologists often knew all the available structures of bio macromolecules and much of the analysis, such as structure comparison and hydrogen-bonding analysis was done manually without much computer intervention. As the number of available structures increased, so did the need for software tools for structure analysis. Structural bioinformatics encompasses a wide variety of applications, including structure visualization, validation, classification, structure-based function prediction, identification of targets for structural genomics projects and prediction of the structures of individual bio macromolecules and complexes as well the binding modes of small molecules inside proteins of pharmaceutical or biotechnological interestStructural and theoretical analyses of proteins are central to the understanding of complex molecular mechanisms and are fundamental to the drug discovery process. Computational techniques yield useful insights into an ever-wider range of bimolecular systems. Protein three-dimensional structures and molecular functions can be predicted in some circumstances, while experimental structures can be analyzed in depth via such computational approaches. Non-covalent binding of bio molecules can be understood by considering structural, thermodynamic and kinetic issues, and theoretical simulations of such events can be attempted. The central role of electrostatic interactions with regard to protein function, structure and stability has been investigated and some electrostatic properties can be modeled theoretically. Structural bioinformatics, like many other sub disciplines within bioinformatics, is characterized by two goals: the creation of general purpose methods for manipulating information about biological macromolecules, and the application of these methods to solving problems in biology and creating new knowledge. These two goals are intricatelylinked because part of the validation of new methods involves their successful use in solving real problems. At the same time, the current challenges in biology demand the development of new methods that can handle the volume of data now available, and the complexity of models that scientists must create to explain these data.

STRUCTURAL BIOINFORMATICS: CATALYZED BY LARGE AMOUNTS OF DATA

Biology has attracted computational scientists over the last 30 years in two distinct ways. First, the increasing availability of sequence data has been attracting those with an interest in analysis, algorithms, and probabilistic models (Gusfield, 1997; Durbin *et al.,* 1998). The major accomplishments in this area have been the development of algorithms for pairwise sequence and multiple alignments, the definition and discovery of sequence motifs, and the use of probabilistic models, such as Hidden Markov Models (HMM) to fish out genes of interest. Second is the increasing availability of structural data for those with an interest in computational geometry, computer graphics, and algorithms for analyzing crystallographic and NMR data and creating molecular models. Structural bioinformatics has its roots in this second group. The elucidation of the structure of DNA in the mid-1950s and the publication of the first protein crystal structures in the early 1960s created a demand for computerized methods for examining these complex molecules. At the same time, the need for computational algorithms to decipher X-ray crystallographic data, and to fit the resulting electron densities to the more manageable ball-and-stick models, created a cadre of structural biologists who were very well versed in computational technologies.

GOALS

Advances in structural bioinformatics and availability of high resolution structural information about biological systems will allow us to reason precisely about the function of biological systems. Structural biological analyses have goals that will offer an ultimate insight into the mechanisms of these systems, and therefore a more profound understanding of how biological function follows from structure. The goals of structural bioinformatics lies in four areas: (1) creating an infrastructure for building up structural models; (2) ability to understand the design principles of proteins so that new functionalities can be created; (3) learning how to design drugs efficiently based on structural knowledge of their target; and (4) catalyzing the development of simulation models that can give insight into function based on structural simulations. Each of these four areas has already proved to be successful, and the structural genomics projects promise to create data sets sufficient to catalyze accelerated progress in all these areas.

SPECIAL CHALLENGES FOR STRUCTURAL BIOINFORMATICS

Structural bioinformatics must overcome some special challenges that are either not present or not dominant in other types of bioinformatics domains (such as the analysisof sequence or microarray data). Some of them are:

1. Structural data is not linear and is not easily read to algorithms based on strings. Apart form that, there are also nonlinear relationships between atoms (the forces are not linear), which means that most computations on structure need either to make approximations or to be very expensive.

2. There is a fundamental connection between molecular structure and physics. Although this statement seems obvious and trivial, it means that when reduced representations, such as pseudo atoms or lattice models are applied, they become more difficult to relate to the underlying physics that govern the interactions.

3. Structural data, like all biological data, can be imperfect. Despite some successes in the elucidation of very high-resolution structures, the precision of our knowledge about many structures is likely to be limited by their flexibility and dynamics. Understanding the protein structural disorder may be critical for understanding the protein's function.

4. Protein or nucleic acid structures are generally conserved more than their associated sequence. Thus, sequences will accumulate mutations over time that may make identification of their similarities more difficult, while their structures may remain essentially identical. However, sequence information is still much more abundant than structural information, and so for many molecules it is the sequence information that is readily available. Thus, the need to identify distant sequential similarities in order to gain structural insights can be a major challenge.

5. It must recognize that there is a major gap in our knowledge of a large fraction of proteins that are not globular and water soluble. In particular, membrane-bound and fibrous proteins are simply not well understood and structures are not available in the large numbers. The importance of this shortcoming cannot be overemphasized, since these classes of proteins

are among the most important for understanding a large number of cellular processes of great interest, including signal transduction, cytoskeletal dynamics, and cellular localizations and compartmentalization.

CONCLUSION

Structural analyses of proteins are crucial to understand fundamentals of molecular structures and their role in drug discovery process. Computational techniques have provided us useful insights into biomolecular systems. Protein three-dimensional structures and molecular functions can be predicted in some circumstances, while experimental structures can be analyzed in depth via such computational approaches. Structural bioinformatics is proving to be powerful tools for the future biologists once if all the challenges in its path are dealt with properly. The challenges are only a few in the field of Structural Bioinformatics.

REFERENCES

Altman, R.B. 1998. A curriculum for bioinformatics: the time is ripe. *Bioinformatics,* 14:549-50.

Altman, R.B. and Raychaudhuri S 2001. Whole-genome expression analysis: challenges beyond clustering. *Curr Opin Struct Biol.,* 11(3):340-7.

Baker, D. and Sali, A. 2001. Protein structure prediction and structural genomics. *Science,* 294:93-96.

Bourne, Philip, E. II, Weissig Helge. 2003. Defining bioinformatics and Structural bioinformatics. *Structural Bioinformatics,* 4:3-8.

❑❑❑

11

Database for Animal Genetic Resources of A & N Islands

M. Balakrishnan, S. Jeyakumar, R.C. Srivastava, Mayank Pokhriyal and Soma Mondal

Biodiversity conservation and sustainable use has, for many people, become associated mainly with issues related to plants and wild animals. Although much less discussed, the loss of farm animal genetic resources may well be much more serious than in crops because the gene pool is smaller and very few wild relatives remain. The fact that 32% of livestock breeds worldwide are at risk of becoming extinct and that the rate of extinction continues to accelerate (FAO, 2000) is thus a serious cause for concern. Animals of different characteristics and hence outputs suit differing local community needs.

The union territory of A & N Islands constitutes the farthest and the remotest part of Indian union. The islands are situated in the Bay of Bengal surrounded by sea. The neighboring countries are Burma in north east and Singapore, Malaysia and Indonesia in south east. The Union territory of A & N Islands is situated between 6^0 and 14^0 north latitude and 92^0 and 94^0 east longitude forming a broken row of continuous islands from. North to South direction. They can be divided into two groups namely Andaman and Nicobar groups. The dreaded 10^0 channel which is about 145 Kms wide and 400 fathoms deep, separates the two groups

It is needless to state that in a predominated agricultural economy like our nationanimal genetic resources play a vital role. Livestock products no doubt find a place in the list of essential commodities in the day to day life common man.

Unlike the mainland limited variety of livestock are found in these islands. Many domestic animals like Horses, Mules, Ponies, sheep etc are not found in these islands. The cattle, Buffaloes,Goats, Pigs and Poultry are the important livestock present in these islands. The loss of livestock diversity seriously reduces our potential to alleviate poverty, improve food security and promote sustainable agriculture. It has become extremely essential to document the animal genetic resource on these islands. Keeping this in mind a database depicting the animal genetic resources of A & N Islands namely AGRANI has been created.

DATABASE OVERVIEW AND FUNCTIONAL SCOPE

The database is aimed at developing the website - AGRANI – Animal Genetic Resources of A & N Islands. The website will prove to be a databank of animal genetic resources of these islands and will contain information like.....the AGRANI website will provides the administrator user to create users with permissions to access the data of their respective area of work. The website is targeted to a wide range of user which includes department of Animal Husbandry and Veterinary Services, policy makers, Scientists and public.

LAYOUT

The database in designed in Access 2003 at the backend and the interface is created using Adobe Dream weaver in HTML format. The webpage interface is interactive and is designed keeping general public in mind for easy access of their desired data. The database is very rich and up-to-date with information from latest Island wise Statistical census and other data from animal husbandry, CARI and other Veterinary Services.

The main page of the database comprises of links to WebPages of CARI, A & N Islands, Veterinary resources *etc*. To access the database one needs to login to the website. Once logged on the user will get access to the entire database regarding different animal genetic resources in these islands. The AGRANI database is rich in information. The databases comprises of general information about A & N Islands, fields for livestock, poultry and wild or captive animals. The databases will give a complete information regarding family, genus, species, chromosome no, location, breed and community rearing the breed.

CONCLUSION

It is needless to state that animal genetic resources play an essential role in agricultural economy of our nation. Livestock products have a vital place in the list of essential commodities in the day to day life common man. When compared to mainland, very limited variety of livestock is found in these islands. Some major livestock includes cattle, goat, buffaloes and pigs. We need to tap this potential resource to our benefit. This database provides a common place for all the information a concerned (Scientist or common man) person need for his benefit. These databases would be updated on a regular basis so that it could prove useful to Scientists, policy makers and common man.

REFERENCES

Barker, J.S.F. 1980. Animal genetic resources in Asia and Oceania – The perspective, 13-19 in Proc. SABRAO Workshop on Animal Genetic Resources in Asia and Oceania. Tropical Agriculture Research Center, Tsukuba, Japan.

Baumung, R., Simianer, H. and Hoffmann, I. 2004. Genetic diversity studies in farm animals – a survey. *Journal of Animal Breeding and Genetics*, 121(6):361-373.

FAO. 1984. Animal genetic resources conservation by management, data banks and training. FAO Animal Production and Health Paper 44/1. FAO, Rome. 186.

Ruane. 1999. A critical review of the value of genetic distance studies in conservation of animal genetic resources. *Journal of Animal Breeding and Genetics*, 116(5):317-323.

12

A Database for Medicinal Plants used in Treatment of Asthma

K. Balaji, M. Rajadurai, G. Vidhya, A. Krithika, E. Hudson, Vidya Murali, M. Ramya, Anusha Bhaskar and V. Stanley

Among several respiratory diseases affecting man, bronchial asthma is the most common disabling syndrome, nearly 7 to 10 per cent of the world population suffer from bronchial asthma, which is a chronic illness involving the respiratory system in which the airway occasionally constricts, becomes inflamed, and is lined with excessive amounts of mucus, often in response to one or more triggers. These episodes may be triggered by such things as exposure to an environmental stimulant (or allergen), cold air, warm air, moist air, exercise or exertion, or emotional stress. In children, the most common triggers are viral illnesses such as those that cause the common cold. A variety of exogenous determinants of asthma were also been, some of which are air pollution, tobacco smoking, diet, occupation and respiratory infection, these pollutants may serve as inciters or triggers of asthmatic reaction in airways that are already hyper responsive, they may exert direct toxic influence on the respiratory epithelium or may augment or modify immune responses to inhaled antigens. During an asthma episode, inflamed airways react to environmental triggers such as smoke, dust, or pollen. The airways narrow and produce excess mucus, making it difficult to breathe. In essence, asthma is the result of an immune response in the bronchial airways. As according to the World Health Organization over 80% of the people in developing countries depend upon traditional medicine for their primary health care. Medicinal plants are considered as source of various alkaloids and other chemical substances essential for mankind. The study of medicinal plants is being revived again. Although they have been used for millennia by tribal and ethnic communities throughout the world, only in recent years medicinal plants have attracted global interest as they constitute a rich treasure trove of cultural information and are sources of natural products, which provide health security to millions in rural communities. In recent years there has been renewed interest in natural medicinesthat are obtained from plant parts or plant extracts. Nearly 40 percent or more of the pharmaceuticals currently used in western countries are derived or at least partially derived from natural sources.Standard bronchodilator drugs are available, and the relief offered by them is mainly symptomatic and short lived. Moreover, the side effects of these drugs are also quite disturbing. Hence, there is a quest for effective and safe remedy for bronchial asthma. In recent years there is a global trend for revival of interest in the traditional system of medicine. Screening of medicinal herbs has become a potential source of biodynamic compounds of therapeutic value in phytochemical research. Hence we created AsthmaPlantBase, an asthma e-literature database of medicinal plants with taxonomical position, abstract, plant parts used, references, other medicinal uses etc.,

METHODOLOGY

Construction of AsthmaPlantBase

Data of plants used in the treatment of asthma were collected from various literature sources such as Pubmed, ScienceDirect, Blackwell Synergy, Scirus, Ingenta Connect, Springerlink, Biomed Centraland from use in folklore medicine. The database includes 40 genus and 50 species of plants having effect in the treatment of asthma. The list however, is not complete. The data is provided in alphabetical order and the records are organized to simplify the task of finding relevant data of any plant. The database can be accessed alphabetically either using genus name or vernacular name for detailed information of the plant.

DATABASE DESIGN

The database is developed using MYSQL as back-end and ASP.net and PHP as front-end on multiple platform and updated regularly. Each entry in AsthmaPlantBase is provided with a unique accession number AXX01.03, where, A denotes asthma, XX denotes genus and species names, 01 represents the alphabetical position, 03 is the entry record. The entry of plants listed in the database contain information such as plant name (Scientific and common), part investigated, active components attributing the medicinal property. The "Article link" provides few free text articles for further reading. A screen shot of the database is given in figure 1.

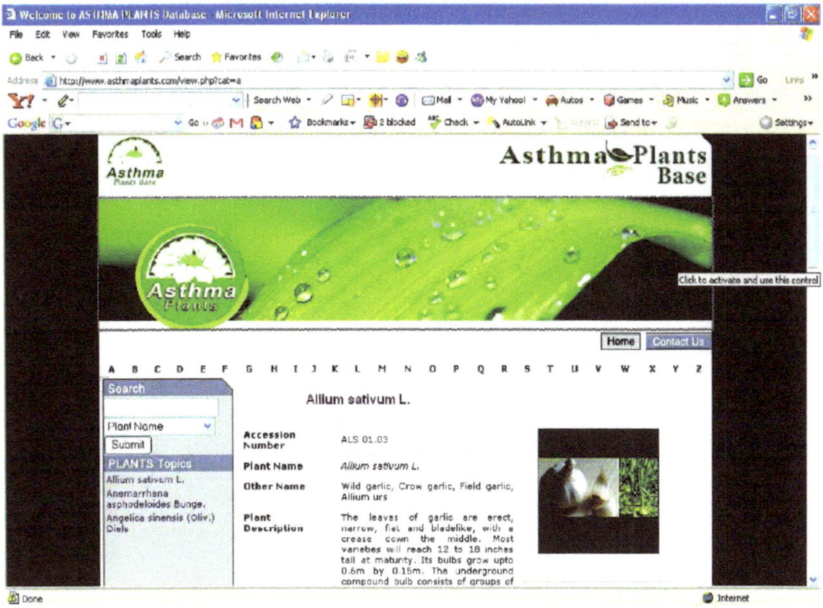

Fig.1: A screen shot of *Allium sativum* record entry in asthma plant base is displayed.

UTILITY

AsthmaPlantBase enlists plants used in the treatment of asthma. This is however, the first database containing medicinal plants for treatment of asthma. The database finds utility to the scientific community for a quick review on the number of plants and plant Parts for asthma medicinal plant research and may serve as a platform for development of drugs.

FUTURE DEVELOPMENT

Currently the database can be accessed on multiple platforms. We also plan to update and refine this database periodically.

REFERENCES

Govindan S, *et al.,* 1999. *J Ethnopharmaco.,*66: 205 [PMID: 10433479].
Kamboj VP. 2000.*Curr. Sci.*, 78: 35.
KimberI. 1998. *Toxicol Lett*, 102: 301 [PMID: 10022270]
Kiran SG*et al.,* 2004. *Ind J Biotechnol.*, 3:103 pp.
Maddox L & Schwartz DA, 2002.*Annu. Rev. Med.,*53:477 [PMID 11818486]
Mudappa A & Oommen S. 1998. *Amruth.*, 2:10
Rout GR, *et al.,* 1991.*Plant Cell Tiss. Org. Cul.*, 27: 65
Zhao J, *et al.,* 2002. *J Pediatr Allergy Immunol.,* 13:47 [PMID 12000498]
http://www.biomedcentral.com
http://www.blackwell-synergy.com

http://www.ingentaconnect.com/content
http://www.ncbi.nlm.nih.gov/entrez/query.fcgi?db=Pubmed
http://www.sciencedirect.com
http://www.scirus.com
http://www.springerlink.com
http://www.who.int/medicinedocs/

❑❑❑

13

In-silico Determination of Potential Drug Target for Curing Meningococcal Disease- Bioinformatics Study On Penicillin- Binding Proteins

Vijay Tripathi, Dwijendra Gupta and Navita Srivastava

Neisseria meningitidis, also simply known as meningococcus, is a gram-negative bacterium best known for its role in meningitis. It only infects humans; there is no animal reservoir. It is the only form of bacterial meningitis known to cause epidemics. The carbohydrate capsule of *N. meningitidis* determines its virulence and is targeted by the immune system. Approximately 12 strains of *N. meningitidis* exist and are characterized by the polysaccharide expressed on its capsule: A, B, C, 29-E, H, I, K, L, W-135, X, Y and Z. Inhibition of Pencillin Binding proteins (PBPs) leads to irregularities in cell wall structure such as elongation, lesions, loss of selective permeability and eventual cell death and lysis. 3-D structure of PBP was studied using Bioinformatics tool. Using Homology modelling, converted protein sequence of penicillin binding protein, extracted from NCBI server, in FASTA format and performed PDB-BLAST to search a desired template. Selection of template was based on lower E-value, bit score and identity. Using MSA with the help of CLUSTAL W software conserved region was found between the target and template sequence, 1K25. The 3-D structure of penicillin binding protein was found with the help of Modeller 9v2 and molecular dynamics was performed using GROMACS software. The docking was performed (Autodock 3.0 software) using 3 ligands *viz.* Cefuroxime, Chloramphenicol, Cefalotoxime with standard Bioinformatics tools.

ENERGY; Differences between the model's features and restraints:

Number of all residues in MODEL	: 582
Number of all, selected real atoms	: 4490 4490
Number of all, selected pseudo atoms	: 0 0
Number of all static, selected restraints	: 0 0
COVALENT_CYS	: F
NONBONDED_SEL_ATOMS	: 1
Number of non-bonded pairs (excluding 1-2,1-3,1-4)	: 918582
Dynamic pairs routine	: 1, NATM x NATM double loop
Atomic shift for contacts update (UPDATE_DYNAMIC)	: 0.390
LENNARD_JONES_SWITCH	: 6.500 7.500
COULOMB_JONES_SWITCH	: 6.500 7.500
RESIDUE_SPAN_RANGE	: 1 9999
NLOGN_USE	: 15
CONTACT_SHELL	: 15.000
DYNAMIC_PAIRS,_SPHERE,_COULOMB,_	
LENNARD,_MODELLER	: T F F F T
SPHERE_STDV	: 0.050
RADII_FACTOR	: 0.820
Current energy	: -44591.9805

Molecular dynamics was studied with the help of GROMACS software and found minimum energy (5.87834e+04) of this model (Fig.1).

Results on the basis of Binding Energy, Docking Energy and maximum Hydrogen Binding suggest that Chloramphenicol is the best drug since it gave best docking with the macromolecule (protein).

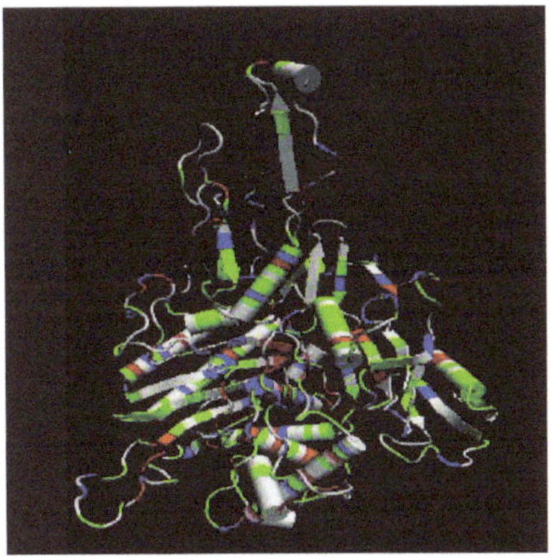

Fig.1:

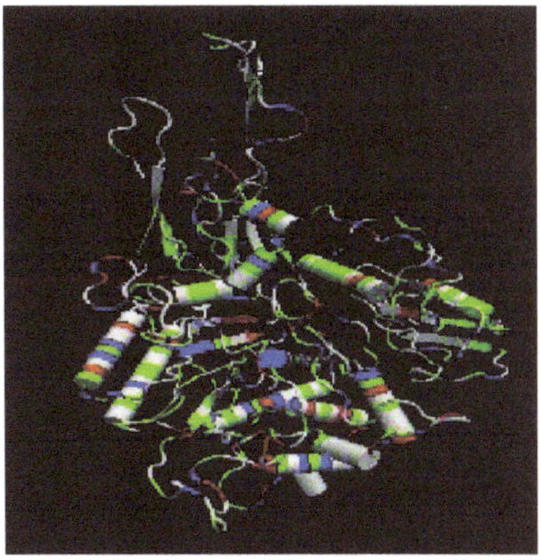

Fig. 2:

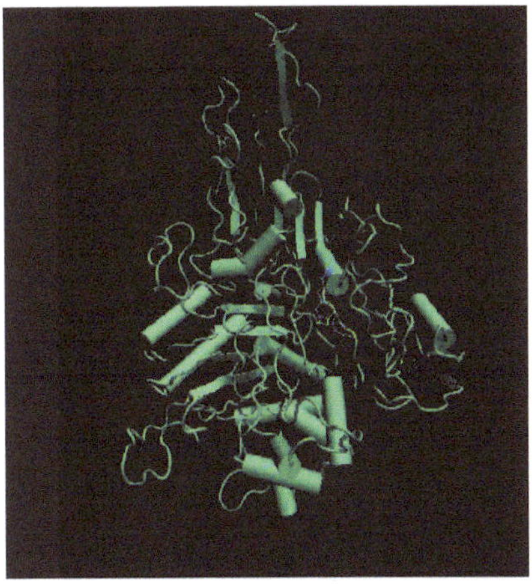

Fig. 3:

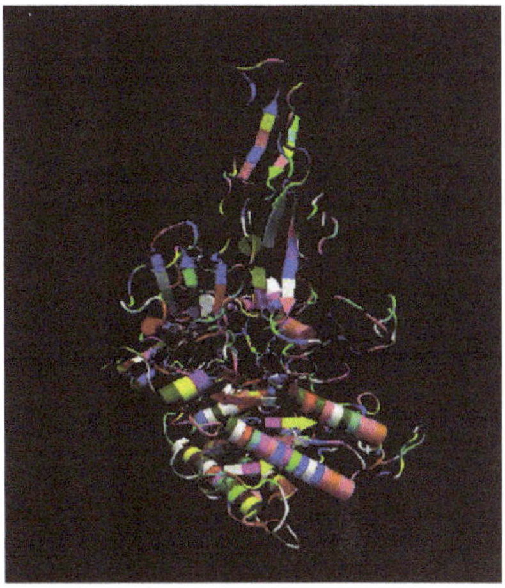

Fig. 4:

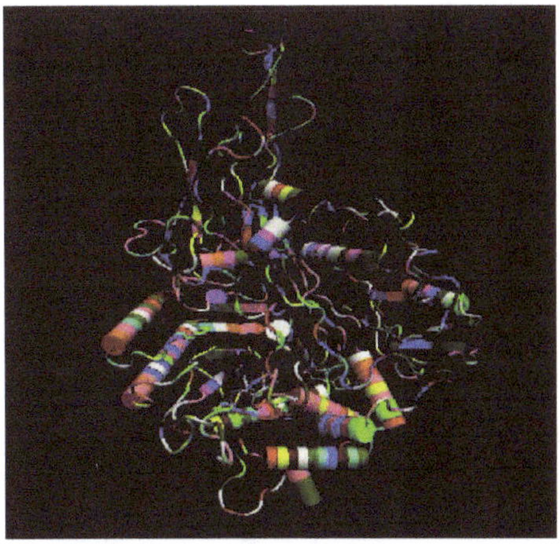

Fig. 5:

Table 1: Docking with Cefuroxime

S.No.	Binding Energy	Docking Energy	Inhibition constant	Internal molecular Energy	Torsional Energy	Internal Energy	d-RMS	Hydrogen bond
1.	-3.96	-6.77	0.0	-6.76	2.8	0.01	81.79	1
2.	-2.93	-6.36	0.01	-5.73	2.8	0.63	101.486	1
3.	-4.35	-7.6	0.000649	-7.15	2.8	0.45	108.774	2
4.	-6.34	-9.34	2.29-e05	-9.14	2.8	0.2	116.001	0
5.	-4.07	-7.5	0.07	-6.87	2.8	0.63	182.653	3
6.	-6.41	-9.38	2e-05	-9.21	2.8	0.17	116.028	0
7.	-4.77	-8.03	0.000319	-7.57	2.8	0.45	109.4	1
8.	-2.98	-6.21	0.01	-5.78	2.8	0.43	105.913	0
9.	-4.06	-7.47	0.0	-6.87	2.8	0.6	82.5	2
10.	-5.98	-6.78	0.0	-6.53	2.8	0.73	112.878	2

Table 2: Docking with Chloramphenicol

S.No.	Binding Energy	Docking Energy	Inhibition constant	Internal Molecular Energy	Torsional Energy	Internal Energy	d-RMS	Hydro-gen bond
1.	-4.78	-7.28	0.000135	-6.96	2.18	0.33	98.01	2
2.	-3.68	-6.3	0.0	-5.86	2.18	0.44	110.98	3
3.	-4.54	-7.14	0.000472	-6.72	2.18	0.42	98.62	3
4.	-3.91	-6.49	0.0	-6.09	2.18	0.41	102.92	3
5.	-4.73	-7.3	0.00034	-6.91	2.18	0.39	98.60	2
6.	-4.71	-7.29	0.000351	-6.89	2.18	0.39	98.06	2
7.	-4.86	7.4	0.000273	-7.04	2.18	0.36	110.93	1
8.	-4.03	6.35	0.0	-6.21	2.18	0.13	110.01	1
9.	-4.0	6.51	0.0	-6.18	2.18	0.33	99.05	4
10.	-5.98	6.78	0.0	-6.53	2.8	0.73	112.88	3

Table 3: Docking with Cefalotoxime

S.No.	Binding Energy	Docking Energy	Inhibition constant	Internalmolecular Energy	Torsional Energy	Internal Energy	d-RMS	Hydro-gen bond
1.	-0.81	-3.92	0.26	-3.92	3.11	0.0	128.01	1
2.	-2.22	-5.33	0.02	-5.33	3.11	0.0	81.9	2
3.	-1.19	-4.3	0.13	-4.3	3.11	0.0	127.25	1
4.	-1.94	-5.05	0.04	-5.05	3.11	0.0	108.59	1
5.	-0.05	-3.07	0.0	-3.07	3.11	0.0	110.9	1
6.	-1.39	-4.5	0.1	-4.5	3.11	0.0	126.19	1
7.	-0.98	-4.09	0.19	-4.09	3.11	0.0	86.85	1
8.	-0.61	-3.73	0.36	-3.73	3.11	0.0	96.31	0
9.	-1.44	-4.55	0.09	-4.55	3.11	0.0	105.36	0
10.	-0.61	-3.73	0.36	-3.73	3.11	0.0	132.96	2

ACKNOWLEDGEMENT

Supported by a Department of Biotechnology Grant (DO.BT/BI/04/055/2001) for creation of Bioinformatics Infrastructure Facility under BTISNet scheme to University of Allahabad.

14

Text Mining: Potential Applications in Bioinformatics

M. Balakrishnan, R.C. Srivastava, S.K. Zamir Ahmed and Subhash Chand

A fundamental problem facing researchers today is how to make effective use of available wealth of online background domain knowledge to improve their understanding of complex biological system. Effective use of such online background domain knowledge plays a crucial role in all stages of integrated biological studies.

In the past years, research in molecular biology and molecular medicine has accumulated enormous amounts of data. This includes genomic sequences gathered by the Human Genome Project, gene expression data from micro array experiments, protein identification and quantification data from proteomics experiments, and SNP data from high-throughput SNP arrays. However, the understanding of the biological processes underlying these data lags behind. There is a strong interest in employing methods of knowledge discovery and data mining to generate models of biological systems. Mining biological databases put forward challenges which knowledge discovery and data mining have to address in the future. Important background knowledge in bioinformatics is often contained in documents, such as scientific publications or database annotations. Analyzing data from biological databases often requires the consideration of data from multiple relations rather than from one single table. Recently, approaches (such as propositionalization algorithms) are being studied that utilize multi-relational data and yet meet the efficiency requirements of large-scale data mining problems.

It is difficult and requires profound understanding of both knowledge discovery and computational biology to identify problems and optimization criteria which, when maximized by knowledge discovery algorithms, actually contribute to a better understanding of biological systems. Identification of appropriate knowledge discovery problems and development of evaluation methods for knowledge discovery results are ongoing efforts.A plethora of online resources both in structured (Data tables) and semi structured (such as XML) about genes and their involvement in identified biological systems are available at the disposal, of scientists worldwide we just need to tap this potential knowledge in an effective manner.

TEXT MINING FOR INTERPRETING GENE EXPRESSION ANALYSIS RESULTS

Let us, consider a scientist or researcher is involved in the analysis of gene expression data using traditional data clustering techniques. The result could be a group of co-regulated genes (i.e. genes that exhibit similar experimental behaviour) or could be groups of differentially expressed genes. Once theses groupings are isolated, the scientist can investigate or validate his results by:

i) Looking for information about these differentially expressed or co regulated groups of genes.

ii) Identifying the diseases associated with these groups of genes.

Such information is available at online genomic or microarray expression databases and in scientific publications. What the user needs is an interactive method which will enable him to access the information dynamically, summarize it and then refurnish it in his analysis.

A good example of this is the Discovery net workflow system were the scientist with the help of info grid integration framework will extract information about the co-regulated or differentially expressed genes from the online databases such as NCBI. The retrieved sequence cab be used to execute a BLAST query to retrieve a set of homologous sequences; these sequences in turn can be used as a query to SwissProt database to retrieve the PubMed Ids identifying articles or abstracts associated with these articles (M. Ghanem., *et,al)*. Methods like these can also be used to correlate between different experimental data also.

TEXT MINING FOR PROTEIN STRUCTURE PREDICTION

Every newly sequenced genome is annotated by comparing its sequences to previously annotated genomes. Around 60% of a new genome is reliably annotated in this manner. Once protein's approximate structure is predicted we can compare that structure to proteins of known function. But this approach is less specific and hence it is better to confirm the structure prediction based annotations by other means also.

We can generate textual information for any sequences from sequence-based comparison to the non-redundant UniProt sequence database by selecting the literature associated with the top BLAST hits and extracting MeSH keywords. Similarly, literature can be obtained and combined for the sequences of all structural SCOP super families.

TEXT MINING FOR BIOMEDICAL APPLICATIONS

Biomedical articles provide functional descriptions of bioentities such as chemical compounds and proteins. To extract relevant information using automatic techniques, text-mining and information-extraction approaches have been developed. These technologies have a key role in integrating biomedical information through analysis of scientific literature. Some important applications include the identification of biologically relevant entities in free text and the construction of literature-based networks of protein–protein interactions. Also, text mining can aid the interpretation of microarray data and the analysis of pathology.

Besides the recognition of protein interactions from scientific text, natural language processing has been applied to a broad range of information extraction problems in biology. Hahn *et al.,*(2002) used natural language techniques and nomenclatures of the Unified Medical Language System (UMLS) to learn ontological relations for a medical domain. Baclawski *et al.,*(2000) developed a diagrammatic knowledge representation method called keynets

TEXT MINING AND DATABASE CURATION

Database curation is interesting since curated databases represent a repository of 'gold standard' data. Craven and Kumlien (1999) reported an experiment where they used the subcellular localization field of the Yeast Protein Database (Hodges *et al.,*, 1998). They collected instances of this relation from the database, traced the references associated with each database entry back to the PubMed abstract, and then within each abstract, identified the sentence that gave rise to the annotation. This gave them a set of extracted relations (from the database) and the underlying text sources (sentences from the abstracts). This experiment suggests how curated databases might be exploited to create 'cheap' annotated information.

CONCLUSION

In this paper it is shown how of literature data mining is applied in various areas of bioinformatics and the need for challenges it faces. It shows how current language processing approaches can be successfully used to extract and organize information from the literature. It also illustrates the diversity of these applications. By defining several biologically important challenge problems and by providing the associated infrastructure, we can accelerate progress in this field. This will allow us to compare approaches, to scale up the technology to tackle important problems, and to learn what works and what areas still need work.

ACKNOWLEDGEMENT

Supported by a Department of Biotechnology Grant (DO.BT/BI/04/055/2001) for creation of Bioinformatics Infrastructure Facility under BTISNet scheme to University of Allahabad.

15

Diversity and Antibacterial Activity of Coral Associated Actinomycetes

P. Nithyanand, S. Manju and S. Karutha Pandian

The coral reef ecosystem is one of the most diverse ecosystems, which is housing for an amazing diversity of organisms and estimation of species richness ranges from 650 000 to over 9 million. For this reason, coral reefs often are referred to as the "rainforests of the ocean"[1].Prokaryotes represent overriding life form on the planet and are almost certainly the most diverse component of coral reef communities[2]. Corals have been known to harbour many different species of bacteria.

The rise in the number of drug-resistant pathogens and the limited success of strategies in providing new agents indicate an uncertain forecast for future antimicrobial therapy. So it is critical that new groups of microbes from unexplored habitats be pursued as sources of novel antibiotics. Marine actinomycetes serve as a logical target given the historical significance of their terrestrial counterparts in antibiotic production. Isolation strategies directed towards new marine-derived actinomycetes have been lacking, but it is clear that such procedures will increase the understanding of marine bacterial diversity and yield many novel marine natural products. Though recent literatures report the isolation of novel class of actinomycetes from marine sediments and sponges, reports on actinomycetes associated with corals are very scanty. Therefore in this study we attempted to characterize the actiomycetes diversity associated with the coral *Acropora digitifera* from the Gulf of Mannar, Southern coast of India. The antibacterial potential of the isolated actinomycetes was also investigated.

MATERIALS AND METHODS

Sample Collection

Mucus from the coral, *Acropora digitifera* was collected from Gulf of Mannar, situated in the southern part of the peninsular India which extends from AdamsBridge to Cape Comorin. Mucus samples were collected from the coral by swabbing the surface of the coral using a sterile cotton swab according to Guppy *et al*[5]. The cotton swab was immediately placed in a sterile centrifuge tube and transported back to the laboratory in 4°C.

Isolation of Coral associated actinomycetes

Actinomycetes were isolated and cultivated from the coral by serially diluting the coral mucus and plating them on Starch Casein Agar supplemented with seawater. All the plates were incubated at 28°C for 7 days.

CHARACTERIZATION OF CORAL ASSOCIATED ACTINOMYCETES

Biochemical identification

All the isolated bacteria were identified by performing various biochemical tests according to the Lampert *et al.*

Molecular identification

Genomic DNA was extracted from selected isolates and 16S rDNA was amplified using actinomycetes specific16S rRNA gene primers (forward primer 5' CGCGGCCTATCAGCTTGTTG 3' and reverse primer 5' CCGTACTCCCCA GGCGGGG 3'). Polymerase Chain Reaction was performed with a 50-μl reaction mixture containing 2 μl (10 ng) of DNA as the template, each primer at a concentration of 25 ρmol, 2.75 mM $MgCl_2$, and each deoxynucleotide triphosphate at a concentration of 0.06 μM, as wellas 2 U of *Taq* polymerase and buffer used as recommended by the manufacturer (MBI Fermentas). After the initial denaturation for 4 minat 95°C, there were 40 cycles consisting of denaturation at 95°C for 45 secs, annealing at 72°C for 1 min, and extension at 72°C for 1 min and then a final extension step consisting of 5 minat 72°C; Mastercycler Personal (eppendorf, Germany) was used. The amplification of 16S rDNA was confirmed by running the amplified product in 1% agarose gel in 1X TAE. With an objective of finding the bacterial diversity, all the amplicons representing various isolates were subjected to Amplified Ribosomal DNA Restriction Analysis (ARDRA). The ARDRA profile obtained with *Hin*fl revealed the presence of different polymorphic groups. 16S rDNA amplicons of the polymorphic strains were selected for sequencing. Sequencing reaction was carried out using ABI PRISM 310 Genetic Analyzer (PE Applied Biosystems). For sequencing reaction Big Dye Ready Reaction DyeDeoxy Terminator Cycle Sequencing kit (Perkin-Elmer) was employed. The resulting partial sequences were matched with sequences available in NCBI using BLAST.

Sequences were preliminarily matched with previously published bacterial 16S DNA sequences using BLAST[10]. The sequences were aligned with the most similar sequences and the sequences of other representative bacterial 16S rDNA regions using ClustalW[11], and a phylogenetic analysis was performed by using PHYLIP (DNADIST, NEIGHBOUR, and SEQBOOT)[12].

RESULTS AND DISCUSSION

Enumeration of Coral associated actinomycetes:

After incubation the plates were enumerated to find the total colony forming units (cfu) the total counts are summarized in Table 1. A total of 22 different isolates were isolated based on the colony morphology and colony colour. Total actinomycetes population was high in mucus when compared to sea water and sediment.

Table 1: Total Count Actinomycetes in Coral reef ecosystem

Sl. No	Bacterial Genera	Colony forming units (cfu/ml)		
		Mucus	Seawater	Sediment
1.	Actinomycetes	7.5x103	2.0x102	4.0x102

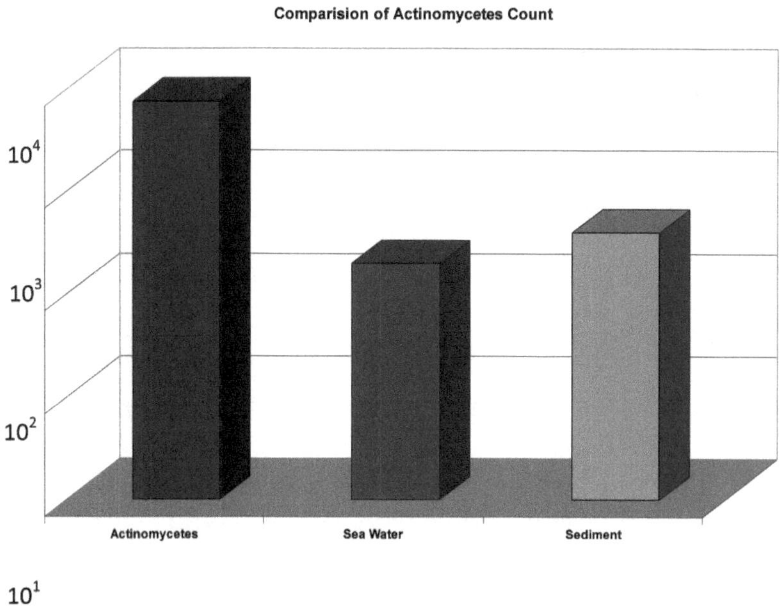

Fig. 1: Comparison of Total Actinomycetes Population in Corals

BIOCHEMICAL IDENTIFICATION OF CORAL ASSOCIATED ACTINOMYCETES

The coral actinomycetes were subjected to various morphological and biochemical tests. Test included Gram staining, Motility, Catalase assay, Indole test, Methyl red test, Voges-Proskauer test, Nitrate Reduction test, Urease, ONPG test, Esculin hydrolysis, Citrate utilization, Malonate utilization and test for utilization of carbohydrates such as Lactose, Xylose, Maltose, Fructose, Dextrose, Galactose, Raffinose, Trehalose, Melibiose, Sucrose, L – Arabinose, Mannose, Inulin, Sodium gluconate, Glycerol, Salicin, Glucosamine, Dulcitol, Inositol, Sorbitol, Mannitol, Adonitol, Alpha –Methyl – D-glucoside, Ribose, Rhamnose, Cellobiose, Melezitose, Alpha – Methyl – D-mannoside, Xylitol, D-Arabinose and Sorbose (Figure 2). Gram staining was performed for all the bacterial isolates. Both Gram positive and Gram-negative bacteria appeared in the total heterotrophic bacteria, which indicate that both classes of bacteria are present in corals.

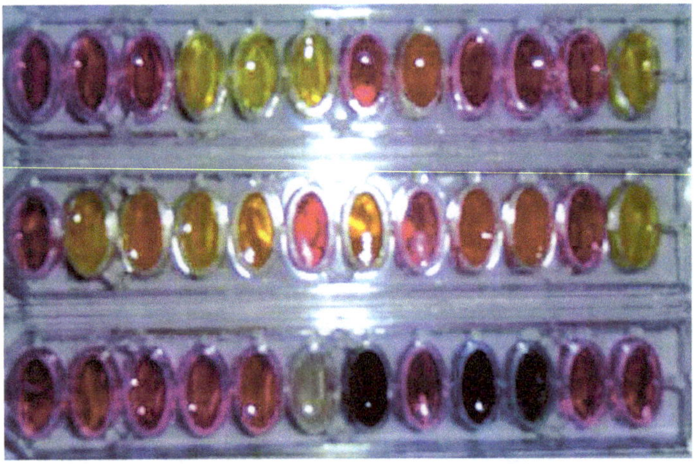

Fig. 2: Different carbohydrates fermented by coral associated actinomycetes

MOLECULAR IDENTIFICATION OF CORAL ASSOCIATED ACTINOMYCETES

Genomic DNA was isolated form all coral associated actinomycetes. DNA samples were visualized on 0.8% agarose gel for PCR analysis. 16S rRNA gene was amplified from all the samples using actinomycetes specific primers. Amplification of 16S rDNA was confirmed by running the amplicons on 1% agarose gel along with 1kb ladder. The size of the amplified fragment of all the samples was approximately 1500bp that corresponded to the prokaryotic 16S rDNA gene (Figure 3).

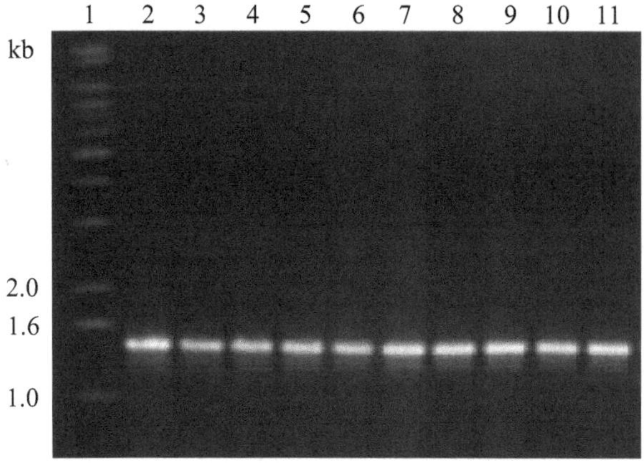

Fig 3: Amplification of 16S rDNA gene from coral bacterial isolates.

SCREENING OF ACTINOMYCETE ISOLATES FOR ANTIBACTERIAL ACTIVITY

Primary Screening: Primary screening was done by T – streak method. All isolates were screened against *Escherichia coli, Bacillus subtilis, Staphylococcus aureus, Salmonella typhi, Vibrio parahaemolyticus, Vibrio vulnificus* and *Vibrio harveyi.*

Secondary Screening: Two potential isolates CA–10 and CA–16 were further screened. CA- 10 and CA -16 were grown on ISP 2 broth for 7 days. After incubation the culture medium was centrifuged and the supernatant was assayed for antibacterial activity. Agar well diffusion method was performed. The culture free supernatant of CA – 10 and CA – 16 showed anti-bacterial activity against *Escherichia coli, Bacillus subtilis, Staphylococcus aureus, Salmonella typhi, Vibrio parahaemolyticus, Vibrio vulnificus* and *Vibrio harveyi*. The diameter (mm) of the growth inhibition halos was measured (Table 2).

Table 2: Antibacterial activity of Coral Actinomycetes Against Different Bacteria

S.N.	Active Isolate	Indicator Strains					
		E. coli	*S. typhi*	*S. aureus*	*B. subtilis*	*V. parahaemolyticus*	*V. vulnificus*
1.	CA - 10	7 mm	11mm	6mm	0mm	12 mm	10mm
2.	CA – 16	5mm	9mm	0mm	13mm	14mm	5mm

Among the two isolates, CA -10 showed high antibacterial activity against aquaculture pathogens like *V. parahaemolyticus* and *V. vulnificus* and enteric pathogen *S. typhimurium* and moderate activity against *E. coli* and *S. aureus.* However it did not show any antibacterial activity against the gram positive bacterium *B. subtilis*. CA – 16 showed high activity against B. subtilis and *V. parahaemolyticus* and no activity against S. aureus. Both CA -10 and CA-16 can be putative probiotic candidates considering their activity against the aquaculture pathogens. Many marine natural products and enzymes have been isolated from Actinomycete genera of marine origin[14]. Presence of *Actiomycetes* in the coral *Acropora digitifera* promises much useful enzymes/bioactive compounds for exploration. Many bioactive compounds such as streptomycin, subtilosin, sublancin and surfactin have already been derived from marine actinomyceteswhich can be used as antibacterial and antifungal agents[14 & 15].Enzymes derived from marine actinomycetes can also be tapped and employed in high salt environment where enzymes from terrestrial bacteria are rendered inactive/ineffective.Moreover, actinomycetes so far have been studied only from marine sponges. The presence of actinomycetes with antibacterial activity in corals opens a new arena in coral microbiology and further augments the field of marine biotechnology in the context of industrial exploration.

ACKNOWLEDGEMENT

The authors thank the Department of Biotechnology (DBT), Ministry of Science and Technology, New Delhi, India for funding the Project (BT/PR3987/AAQ/03/198/2003) and the Bioinformatics Infrastructure Facility, Alagappa University funded by DBT (No. BT/BI/04/055/2001) for the computational facilities provided.

REFERENCES

1. Altschul, S.F., Warren, G., Miller, W., Myers, E.W., Lipman, D.J. 1990. *J. Mol. Biol*, 215:403-410.
2. Coffroth, M.A. 1990. *Mar. Biol*, 105:39-49.
3. Felsenstein.1993. *J. Genet. Res.*, 60:209-220.
4. Frias-Lopez, J., Zerkle, A.L., Bonheyo, G.T., Fouke, B.W. 2002. *Appl. Environ. Microbiol*, 68:2214-2228.
5. Guppy, R., Bythell, J.C. 2006. *Marine Ecology Progress Series*, 328:133-142.
6. Kushmaro, A., Loya, Y., Fine, M., Rosenberg, E. 1996. *Nature*, 380:396.
7. Mohapatra, B.R., Bapuji, M., Sree, A. 2002. *Folia Microbiol (Praha)*,47: 51-5.
8. Nancy Knowlton. 2001. *Proc. Natl. Acad. Sci. USA*, 98: 5419-5425.
9. Patterson, K.L., Porter, J.W., Ritchie, K.B., Polson, S.W., Mueller, E., Peters, E.C., Santavy, D. L, Smith G W. 2002. *Proc. Natl. Acad. Sci. USA*, 99: 8725–8730.
10. Ritchie, K.B., Smith, G.W. 1995. *Mol. Mar. Biol. Biotechnol*, 4:345-354.
11. Rohwer, F., Seguritan, V., Azam, F. and Knowlton, N. 2002. *Mar. Ecol. Prog. Ser.*, 243:1-10.
12. Rosenberg, E., Koren, O., Reshef, L., Efrony, R., Rosenberg, I.Z. 2007. *Nature Reviews Microbiology*, 5:355-362.
13. Shashar, N., Cohen, Y., Loya, Y., Sar, N. 1994. *Mar. Ecol. Prog. Ser.*,111:259–264.
14. Tamehiro, N., Hosoya, Y.O., Okamoto, S., Ubukata, M., Hamada, M., Naganawa, H., Ochil K. 2002. Antimicrob. Agents. *Chemother*, 315-320.
15. Thompson, J.D., Higgins, D.G, Gibson, T.J. 1994. *Nucleic Acids Res*, 22:4673-4680.

16

Decision Tree and its Application in Bioinformatics

Rajni Jain and Amarendra Kumar Mishra

The role of bioinformatics in genomics and molecular biology research can be likened to the role of intelligence gathering in battlefield. Intelligence is clearly very important in leading to a victory in the battlefield. Computational algorithms are the heart of bioinformatics as they enable predictions, understanding and interpretation of the data generated from laboratory experimental research carried by biologist. This article discusses one such approach called Decision tree induction and its application for generating rules from a dataset extracted from biological databases.

Decision Tree (DT) is one of the most popular choices for learning from feature based examples. They are especially attractive in data mining environment for several reasons. First due to their intuitive representation, the resulting classification model is easy to assimilate by humans. Second, DT does not require any parameter setting from the user and thus are especially suited for exploratory knowledge discovery. Third DT can be constructed relatively fast and accuracy of DT is comparable or superior to other classification models.

In the present article, based on the data collected from databases available on Internet, efforts have been made to demonstrate the production of rules for mi-RNA prediction. This chapter is important not only because it attempts to generate rules for mi-RNA prediction, but also because it is an attempt to use data mining algorithms for extracting pattern from real data collected from the available data on the Internet.

The paper is organized as follows. Section 2 introduces Micro RNAs and presents the conventional tools used for miRNA and nc-RNA structure prediction. Section 3 presents a brief overview of Decision tree induction and some references dealing with it. Section 4 presents a case study to demonstrate rule induction using decision tree.

MICRO RNAs

MicroRNAs (miRNAs) constitute a novel, extensive class of small RNAs and play important gene-regulation roles during growth and development in plants and animals. These highly conserved, ~21-mer RNAs regulate the expression of genes by binding to the 3'-untranslated regions (3'-UTR) of specific mRNAs. miRNAs are initially expressed as part of transcripts termed primary miRNAs (pri-miRNAs). They are apparently transcribed by RNA Polymerase II, and include 5' caps and 3' poly(A) tails. The miRNA portion of the pri-miRNA transcript likely forms a hairpin with signals for dsRNA-specific nuclease cleavage. The dsRNA-specific ribonuclease Drosha digests the pri-miRNA in the nuclease to release hairpin, precursor miRNA (pre-miRNA). Pre-miRNAs appear to be approximately 70 bp RNAs with 1–4 nt 3' overhangs, 25–30 bp stems, and relatively small loops. Exportin-5 (Exp5) seems to be responsible for export of pre-miRNAs from the nucleus to the cytoplasm. Exp5 has been shown to bind directly and specifically to correctly processed pre-miRNAs. Once in the cytoplasm, Dicer cleaves the pre-miRNA. Dicer is a member of the RNase III superfamily of bidentate nucleases that has been implicated in RNA interference. The resulting double-stranded RNA has 1–4 nt 3' overhangs at either end. Only one of the two strands is the mature miRNA; some mature miRNAs derive from the leading strand of the pri-miRNA transcript, and with other miRNAs the lagging strand is the mature miRNA. To control the translation of target mRNAs, the double-stranded RNA produced by Dicer must strand separate, and the single-stranded mature miRNA must associate with the RISC. The strand with lower stability base pairing of the 2–4 nt at the 5' end of the duplex preferentially associates with RNA-induced silencing complex (RISC) and thus becomes the active miRNA. Large-scale experimental approaches to miRNA gene finding met with some difficulties in the beginning, illustrated by the fact that these regulators escaped detection for so long. The short length of miRNAs and their ability to act redundantly or to have only a subtle phenotypical impact imposes a limitation to the use of mutagenesis and other conventional genetics techniques (3). Direct cloning, on the other hand, may not detect miRNAs that have very low expression levels or that are expressed only in specific conditions and cell types

MicroRNA GENE DISCOVERY IN PLANTS: BIOINFORMATICS

MicroRNAs have been discovered using three basic approaches: direct cloning, forward genetics, and bioinformatics predictions followed by experimental validation. Several hundred miRNAs have been cloned and sequenced from mouse, human, Drosophila, *C. elegans*, and Arabidopsis. Estimates suggest that 200–300 unique miRNA genes are present in the genomes of humans and mice. The sequences of many of the miRNAs are homologous among organisms, suggesting that miRNAs represent a relatively old and important regulatory pathway. Plant miRNAs primarily have been discovered through direct cloning and sequencing of small cellular RNAs or through comparative genomics. Only a handful of mutations at miRNA loci have been identified in genetic screens. Bioinformatics approaches have identified targets for nearly all plant miRNAs. Several experimental methods have been used to confirm miRNA-target interactions and explore the biological significance of miRNA-mediated regulation. Plant miRNAs are high-level regulators of gene expression that affect numerous aspects of plant biology, especially developmental patterning. Mutants impaired in miRNA biogenesis exhibit severe, pleiotropic abnormalities, and plants that over express particular miRNAs or express miRNA-resistant versions of particular miRNA targets exhibit a wide array of unusual phenotypes.

All maturation steps of plant miRNAs are processed by Dicer-like proteins. The predicted miRNA precursors in plants are much more variable in size than those of animals, ranging from around 60 to a few hundred nucleotides, whereas those in animals are typically~ 70-nt long. In both plants and animals, cloning was the initial means of large-scale miRNA discovery. However, cloning is biased toward RNAs that are expressed highly and broadly. MicroRNAs expressed at low levels or only in specific cell types or in response to certain environmental stimuli are more difficult to clone. Sequence-based biases in cloning procedures might also cause certain miRNAs to be missed. Because of these limitations, bioinformatics approaches to identify miRNAs have provided a useful complement to cloning. A straightforward use of bioinformatics has been to find homologs of known miRNAs, both within the same genome and in the genomes of other species.

A more difficult challenge is to identify miRNAs unrelated to previously known miRNAs. This was first accomplished for vertebrate, nematode, and fly miRNAs, using algorithms that search for conservation of sequence and secondary structure (i.e., miRNA stem-loop precursors) between species, searching for patterns that are characteristic of miRNAs. Although these methods identified numerous potential animal miRNAs, many of which were subsequently confirmed experimentally, they have not been directly useful in finding plant miRNAs because of the longer and more heterogeneous secondary structures of plant miRNA stem-loops. To address this need, several groups devised bioinformatics approaches specific to plant miRNA identification.

Like algorithms for identifying animal miRNAs, these approaches use conservation of secondary structure as a filter, but are necessarily more relaxed in terms of allowed structures. Some approaches take advantage of the high complementarity of plant miRNAs to target messages, implementing the requirement that the candidate has conserved complementarity to mRNAs. This additional filter has been useful for distinguishing authentic plant miRNAs from false positives and recently has been extended to mammalian miRNA gene prediction

CONVENTIONAL STRUCTURE PREDICTION TOOLS FOR miRNA AND NCRNA

Alifold: Alifold predicts consensus secondary structures for the set of aligned RNA and DNA sequences. It is offered by the Vienna RNA package. To each possible consensus secondary structure a mean energy averaged value is assigned in the alignment and bonus energies for compensatory and consistent mutations. A penalty score is given for pair that cannot be formed by all sequences. The input file containing aligns sequences must be in clustalW format. The output file presents results in three sections:

i) Postscript drawing of the predicted structure

ii) Postscript dot plot

The mountain plot represents each pair (i, j) of the optimal structure by a slab ranging from i to j with height proportional to its probability. Helices appear as uphill and downhill slopes. Plateaus indicate unpaired regions. Different colors indicate the different types of base pairs. Non- compatible sequences are represented by shades of gray.

iii) Text output with detailed information about each base pair and colored mountain plot in postscript.

Evofold: EvoFold is a comparative method for identifying functional RNA structures in multiple-sequence alignments. It is based on a probabilistic model-construction called phylo-SCFG and exploits the characteristic differences of the substitution process in stem-pairing and unpaired regions to make its predictions. The input of EvoFold consists of a set of multiple alignments and a phylogenetic tree. The phylogenetic tree must use the same sequence names as the multiple alignments, and it should be estimated from neutrally evolving DNA. The input alignments are accepted in the ama/maf format and the default alignment file name is 'input.ama'. The phylogenetic tree must be given in the newick format. Regarding output, the default EvoFold control file contains two analysis sections where first section pertaining to the fRNA model and second section to the background model.

qRNA: This tool is based on comparative genome sequence analysis. And it detects conserved RNA secondary structures, including both ncRNA genes and cis-regulatory RNA structures.

Xrate:Xrate is an interpreter for phylo-grammars (it is a stochastic grammar whose terminals are alignment columns, typically generated by evolving a continuous-time Markov process on a phylogenetic tree. It is used to discover patterns of evolutionary conservation in sequence alignments. Xrate brings together in one package a lot of the capabilities of several other software tools for molecular evolutionary analysis. A grammar can be "trained" on a database of multiple alignments for measurement of evolutionary rates. Alternatively (or additionally), the grammar can be used to annotate alignments with predicted.

Multiple sequence alignments for statistical detection of RNA secondary structure (MSARi):MSARi is a program for detecting conservation of RNA secondary structure. It searches orthologous nucleotide sequences for statistically significant variations conserving a candidate secondary structure. This approach is growing increasingly feasible as the genomes of ever more organisms are sequenced. MSRAI uses a method for coping with sequence redundancy that is likely to have applications in a large set of other comparison-based search methods.

DECISION TREE

DT is a hierarchical structure with root node at the top and the leaves at the bottom. Following a top down approach, Decision Tree (DT) is constructed recursively by splitting the given set of examples. In DT, the target concept is more compact and is easily interpreted. For DT induction two algorithms by Quinlan, ID3 for discrete attributes and its successor C4.5 for numeric continuous attributes are widely used. A DT may be mapped to the set of decision rules by traversing all the paths from the root towards leaves. For algorithmic details of DT induction reader may see the references.

Conventionally, researchers use regression analysis to determine the contribution of a factor in making a decision. There are two limitations of such approaches. Firstly, applications of the statistical methods are based on the assumptions that condition variables are independent of each other. Secondly, resulting model is represented in the form of coefficients which requires mathematical expertise to understand the resulting model. On the other hand, models induced using data mining algorithms are easy to understand and do not require data mining expertise in understanding the resulting classifier.

CASE STUDY: DECISION TREE FOR DISCOVERING RULES

Dataset: Data regarding Pre-miRNA of Anopheles was compiled from various databanks available on Internet using many tools. The data is finally converted to 16 attributes which are briefly explained in Table 1. Class attribute is true or false and shows the status regarding whether the mi-RNA will lead to production of mature RNA. The objective is to predict the class attributes value on the basis of other attributes presented in the Table 1.The *true* value of the class variable means that production of mature RNA is predicted while false value of the class variable means that production of mature RNA is not predicted.

Table 1: Description of attributes from pre miRNA secondary structure of *Anopheles gambiae*

Attribute ID	Description
ID	Name of the Pre- miRNA
LEN	Length of nuclear tides
NBP	Number of base pairs
BLR	Base pair length ratio (NBP/LEN)
NHP	No. of hair pins (if it is zero, it is not a valid data for miRNA)
HPL	HPL in terms of number of nucleotides
FE	Negative Free Energy (Total free energy required to scale the structure, it is always negative hence the negative sign is ignored)
FEPN	Free Energy Per Nucleotide = FE/LEN
AUC (%)	Percentage of Adenine -Uracil Content in the sequence = (content(A+U)/LEN) *100
MSK	Maximum Stack used for base pair i.e. maximum number of continuous base pairs
SD	Symmetric Difference
MBL	Maximum Bulge Length (when it is not a hair pin then it may be bulge)
MBS	Maximum Bulge Symmetry
MTL	Maximum Tail Length
NTL	Tail Length (0, 1 or 2)
CLASS	True or False (1 for true, 0 for false)

METHOD

The purpose of the experiment is to obtain a simple and easy to understand model to identify the factors that can predict the class attribute based on other attributes.

1. Data were pre-processed for cleaning, feature selection and missing values. The cases having missing values were eliminated from the dataset before further processing. The final dataset was having 136 cases where nearly 50 per cent wee having class as *true* while other 50 per cent were having the class value as *false*.

2. Standard method of predicting the accuracy of a learning technique given a fixed sample of data is m x n Cross-Validation (CV). In CV, data set is partitioned randomly to a predefined fixed number of folds, say *n*. Then each fold in turn is used for testing while the remainder (n-1) folds are used for training. This procedure is repeated n times so that at the end every instance has been used exactly once for testing. This is called n-fold CV. To avoid random errors occurring due to data splitting, the procedure is repeated m times by choosing different random number each time for splitting. It has been observed that many authors recommend the 10X10 CV as the method for the best estimate the performance parameters. Hence 10x10 CV was used for experiments in this paper.

3. For implementation of decision tree method, a popular decision tree algorithm called C4.5 by Quinlan was selected for experiments. Java based implementations in WEKA software were used for experimentation in this paper. WEKA is open source software and is freely available on internet for academic and research purpose. Various parameters required for each algorithm were set as per the default values set in WEKA.
4. For evaluation purpose, accuracy and ability to comprehend the model were considered to be of prime importance. Accuracy is defined as the fraction of test cases, which are classified correctly by the induced model. Test cases refer to the data, which are unknown during the model development phase. Number of attributes in the induced model was treated as measure of simplicity and understanding ability of the model.
5. The best model was selected and the corresponding rules were presented to the domain experts. The model, which is acceptable and validated by the experience of the domain experts, is presented in the next section.

Figure 1: Tree 1

FEPN <= 0.49
 | MSK <= 6: 0 (5.0)
 | MSK > 6: 1 (66.0)
FEPN > 0.49: 0 (61.0)

Table 2: Rules corresponding to tree 1

Rules	Coverage%	Accuracy%
If FEPN<=0.49 and MSK <=6 then class=0	3.7	100
If FEPN<=0.49 and MSK >6 then class=1	50	100
If FEPN >0.49 then class=0	46.2	100

RESULTS AND DISCUSSIONS

The dataset after preprocessing contained 132 instances. Experiments were done using many combinations of input attributes and the results are explained here. 10 fold cross validation scheme was used for evaluating test accuracy. Different trees obtained from different sets of features are illustrated by mapping the corresponding trees to rules (Figures 1-8, Table 2-9). Rules are obtained by following the path from tree to leaves. Coverage of a rule means fraction of total instances which are covered by a rule. Accuracy of a rule means fraction of cases predicted correctly within the coverage of a rule. Features before building a tree has been selected using rough set based techniques. The discussion about rough sets is beyond the scope of the present context.

The Evaluation parameters from these trees are compared in Table 2. Depending on the requirement of the users, prediction rules may be selected. These rules are easy to understand and can be validated easily by matching them with the cases whose class values are known but were not used for building the tree. The column accuracy in the Table 10 represents prediction or validated accuracy.

Overall Correctly Classified Instances: 98.5%

Figure 2: Tree 2	Table3: Rules corresponding to tree 2		
	Rules	Coverage%	Accuracy%
MSK <= 8			
HPL <= 5: 0 (57.0)	If MSK<=8 and HPL <=5 then class=0	43	100
HPL > 5	If HPL > 5 and MSK<=6 then Class =0	6	100
MSK <= 6: 0 (8.0)	If MSK between 6 and 8 and	6.89	90
MSK > 6: 1 (9.0/1.0)	HPL > 5 then Class =1		
MSK > 8: 1 (58.0)	If MSK>8 then Class=1	43.9	1100

Correctly Classified Instances: 97.7%

Figure 3: Tree 3	Table 4: Rules corresponding to tree 3		
	Rules	Coverage%	Accuracy%
FE<=48.75:1 (65.0/1.0)			
FE > 48.75	If FE<=48.75 then class=1	49.24	98.5
FEPN <= 0.44	If FE>48.75 and FEPN<=0.44 and	1.5	100
	NBP<=39 then class=0		
NBP <= 39: 0 (2.0)	If FE>48.75 and FEPN<=0.44 and		
	NBP> 39 then class=1	1.5	100
NBP > 39: 1 (2.0)	If FE>48.75 and FEPN> 0.44	47.7	100
FEPN > 0.44: 0 (63.0)	then class=0		

Correctly Classified Instances 128 96.9697 %

Figure 4: Tree 4	Table 5: Rules corresponding to tree 4		
	Rules	Coverage%	Accuracy%
FE <= 48.75: 1 (65.0/1.0			
| FE > 48.75| LEN <=	If FE<=48.75 then class=1	49.24	98.5
120: 0 (62.0)	If FE>48.75 and LEN<=120 then class=0	46.9	100
| LEN > 120	If FE>48.75 and LEN>120 and		
	MSK <=9 then class=0	2.27	100
| MSK <= 9: 0 (3.0)	If FE>48.75 and LEN>120 and	1.51	100
| MSK > 9: 1 (2.0)	MSK >9 then class=1		

Correctly Classified Instances 96.9697 %

Figure 5: Tree 5	Table 6: Rules corresponding to tree 5		
	Rules	Coverage%	Accuracy%
MSK<= 8|NBP<= 30:0	If MSK<=8 and NBP<=30 then class=0	51.51	97.14
(68.0/2.0)|NBP>30:1(6.0)	If MSK<=8 and NBP>30 then class=1	4.54	100
MSK > 8: 1 (58.0)	If MSK>8 then class=1	43.93	100

Correctly Classified Instances 95%

<table>
<tr><th>Figure 6: Tree 6</th><th colspan="3">Table 7: Rules corresponding to tree 6</th></tr>
<tr><th></th><th>Rules</th><th>Coverage%</th><th>Accuracy%</th></tr>
<tr><td>NHP <= 0: 0 (50.0)</td><td></td><td></td><td></td></tr>
<tr><td>NHP > 0</td><td>If NHP<=0 then class=0</td><td>37.87</td><td>100</td></tr>
<tr><td>| NBP <= 29|</td><td>If NHP>0 and NBP<=29 and</td><td>3.78</td><td>100</td></tr>
<tr><td>| LEN <= 77: 1 (5.0)</td><td>Len<=77 then class=1</td><td></td><td></td></tr>
<tr><td>| | LEN > 77</td><td>If NHP>0 and NBP<=29 and Len>77 and</td><td>3.0</td><td>100</td></tr>
<tr><td>| | | NBP <= 27: 0 (14.0)</td><td>NBP <=27 then class=0</td><td></td><td></td></tr>
<tr><td>| | | NBP > 27: 1 (6.0/2.0)</td><td>If NHP>0 and NBP<=29 and Len>77</td><td>4.5</td><td>475</td></tr>
<tr><td></td><td>and NBP >27 then class=1</td><td></td><td></td></tr>
<tr><td>| NBP > 29: 1 (57.0)</td><td>If NHP>0 and NBP>29 then class=1</td><td>43.18</td><td>100</td></tr>
</table>

Correctly Classified Instances 92.4242 %

<table>
<tr><th>Figure 7: Tree 7</th><th colspan="3">Table 8: Rules corresponding to tree 7</th></tr>
<tr><th></th><th>Rules</th><th>Coverage%</th><th>Accuracy%</th></tr>
<tr><td>BLR <= 0.3|</td><td></td><td></td><td></td></tr>
<tr><td>HPL <= 13: 0 (65.0/2.0)</td><td>If BLR <=0.3 and HPL <= 13 then class=0</td><td>50.75</td><td>97.01</td></tr>
<tr><td>| HPL > 13: 1 (4.0)</td><td>If BLR <=0.3 and HPL <= 13 then class=1</td><td>3.03</td><td>100</td></tr>
<tr><td>| BLR > 0.3| NBP <= 28</td><td>If BLR >0.3 and NBP<=28 and</td><td>2.27</td><td>100</td></tr>
<tr><td></td><td>MBS<=3 then class=1</td><td></td><td></td></tr>
<tr><td>| | MBS <= 3: 1 (3.0)</td><td>If BLR >0.3 and NBP<=28 and</td><td>3.79</td><td>80</td></tr>
<tr><td></td><td>MBS>3 then class=0</td><td></td><td></td></tr>
<tr><td>| | MBS > 3: 0 (4.0/1.0)</td><td>If BLR >0.3 and NBP>28 then class=1</td><td>42.42</td><td>100</td></tr>
<tr><td>| NBP > 28: 1 (56.0)</td><td></td><td></td><td></td></tr>
</table>

Correctly Classified Instances 91.6667 %

<table>
<tr><th>Figure 8: Tree 8</th><th colspan="3">Table 9: Rules corresponding to tree 8p</th></tr>
<tr><th></th><th>Rules</th><th>Coverage%</th><th>Accuracy%</th></tr>
<tr><td>AUC <= 30.86: 0 (40.0)</td><td>If AUC <= 30.86 then class = 0</td><td>30.30</td><td>100</td></tr>
<tr><td>AUC > 30.86</td><td>If AUC between 30.86 and</td><td>49.24</td><td>100</td></tr>
<tr><td>| AUC <= 66.96: 1 (65.0)</td><td>66.96 then class = 1</td><td></td><td></td></tr>
<tr><td>| AUC>66.96: 0 (27.0/1.0)</td><td>If AUC > 66.96 then class = 0</td><td>21.21</td><td>96.4</td></tr>
</table>

Total Correctly Classified Instances 96.96 %

SUMMARY

This article presented one application of a machine learning technique namely decision tree for prediction of miRNAs from hexapoda group of species. An overview of certain conventional computational techniques used for similar predictions is also presented in this article. The results are encouraging and can be applied to other species. Other machine learning techniques especially Support Vector Machine, Principal Component Analysis can also be very useful for such type of predictions in bioinformatics domain.

Table 10: Comparative Analysis of Decision Trees and the Respective Evaluation Parameters

Tree ID	Attributes with their IDs					
	#	IDs	# of leaves	Size of the tree	#of rules	Accuracy (%)
1.	14	LEN, NBP, BLR, NHP, HPL, FEPN, AUC, MSK, SD, MBL, MBS, MTL, NTL	3	5	3	98.5
2.	9	LEN, NHP, HPL, AUC, MSK, SD, MBL, NTL	4	7	4	97.7
3.	15	LEN, NBP, BLR, NHP, HPL, FE, FEPN, AUC, MSK, SD, MBL, MBS, MTL, NTL	4	7	4	96.9
4.	5	LEN, FE, AUC, MSK	4	7	4	96.9
5.	13	LEN, NBP, BLR, NHP, HPL, AUC, MSK, SD, MBL, MBS, MTL, NTL	3	5	3	95.0
6.	5	LEN, NBP, NHP, HPL	5	9	6	92.4
7.	11	LEN, NBP, BLR, NHP, HPL, SD, MBL, MBS, MTL, NTL	5	9	5	91.7
8.	6	LEN, AUC, SD, MBL, NTL	3	5	2	97.0

REFERENCES

Altuvia,Y., Landgraf, P., Lithwick, G., Elefant, N., Pfeffer, S., Aravin, A., Brownstein, M.J., Tuschl, T. and Margalit, H. 2005. Clustering and conservation patterns of human microRNAs. *Nucleic Acids Res.*, 33:2697-2706.

Chatterjee, R. and Chaudhuri, K. 2006. An approach for the identification of microRNA with an application to Anopheles gambiae. *Acta Biochim. Pol.*, 53:303-309.

Gautheret, D. and Lambert, A. 2001. Direct RNA motif definition and identification from multiple sequence alignments using secondary structure profiles. *J. Mol. Biol.*, 313:1003-1011.

Grad, Y., Aach, J., Hayes, G. D., Reinhart, B.J., Church, G.M., Ruvkun, G. and Kim, J. 2003. Computational and experimental identification of C. elegans microRNAs. *Mol. Cell*, 11:1253-1263.

Legendre, M., Lambert, A. and Gautheret, D. 2005. Profile-based detection of microRNA precursors in animal genomes. *Bioinformatics*, 21:841-845.

Norden-Krichmar, T., Holtz, J., Pasquinelli, A. and Gaasterland, T. 2007. Computational prediction and experimental validation of Ciona intestinalis microRNA genes. *BMC Genomics*, 8, 445.

Quinlan, J. R., C4.5: Programs for Machine Learning, Morgan Kauffman 1993.

Skowron, A., Komorowski, J., Pawlak, Z. and Polkowski, L., 2002. Rough sets perspective on data and knowledge. Handbook of data mining and knowledge discovery, 134-149, Oxford University Press, Inc. New York, USA, ISBN:0-19-511831-6.

Wang, X., Zhang, J., Li,F., Gu, J., He, T., Zhang, X. and Li,Y. 2005, MicroRNA identification based on sequence and structure alignment. *Bioinformatics*, 21:3610–3614.

Weaver, D., Anzola, J., Evans, J., Reid, J., Reese, J., Childs, K., Zdobnov, E., Samanta, M., Miller, J. and Elsik, C. 2007. Computational and transcriptional evidence for microRNAs in the honey bee genome. *Genome Biol.*, 8: R97.

Weber,M.J. (2005) New human and mouse microRNA genes found by homology search. *FEBS J.*, 272:59–73.

17

Database for Algae Resources in Andaman and Nicobar Islands

M. Balakrishnan

Algae refer to an assemblage of polyphyletic organism that conducts oxygen evolving photosynthesis other than land plants. They are distributed widely throughout India in the sea, in fresh water and in moist situations on land. Algae also form an important component of biodiversity of Andaman and Nicobar Group of Islands. Database for Algae resources of Andaman and Nicobar Islands are created with an aim of providing an information pool to the researchers working in fields related to fresh water and marine water algae in Andaman and Nicobar Group of Islands. About 313 fresh water algal species belonging to 15 families and 57 marine algal species belonging to 18 families has included in this database among which 40 species has been found to be first time reported in India. Algae have a wide application in a variety of Industries such as Food, Fodder, and Fertilizer etc. They are also important sources of Protein, Iodine, Vitamins, Minerals and Substances of Antibiotic Nature.

Structure of this database was developed considering the following features-

1. Uploading specific information on Internet about Algae resource in Andaman and Nicobar Islands for making it available for users from distant places.
2. Storage and Retrieval of information about species of Algae and its eco-geographical distributed areas in the Island.

3. Structure suitable for Storage and updating of molecular data specific to Algae.

ALGAE DATABASE

The Database, Algae Resources in Andaman and Nicobar Islands reveals the geographical distribution of algae species, scientific name, taxonomy, structure, size, locality and its economic importance. The entire database was organized in a object oriented relational database using MS- Access 2003 at the Back end and the interface was created using Adobe Dream weaver in HTML format. Since algae species are widely distributed throughout the Island, this database plays a key role for locating its specific area. Besides this it also contains separate section that deals with genetic information. It helps the researchers to understand the molecular makeup of these polyphyletic organisms.

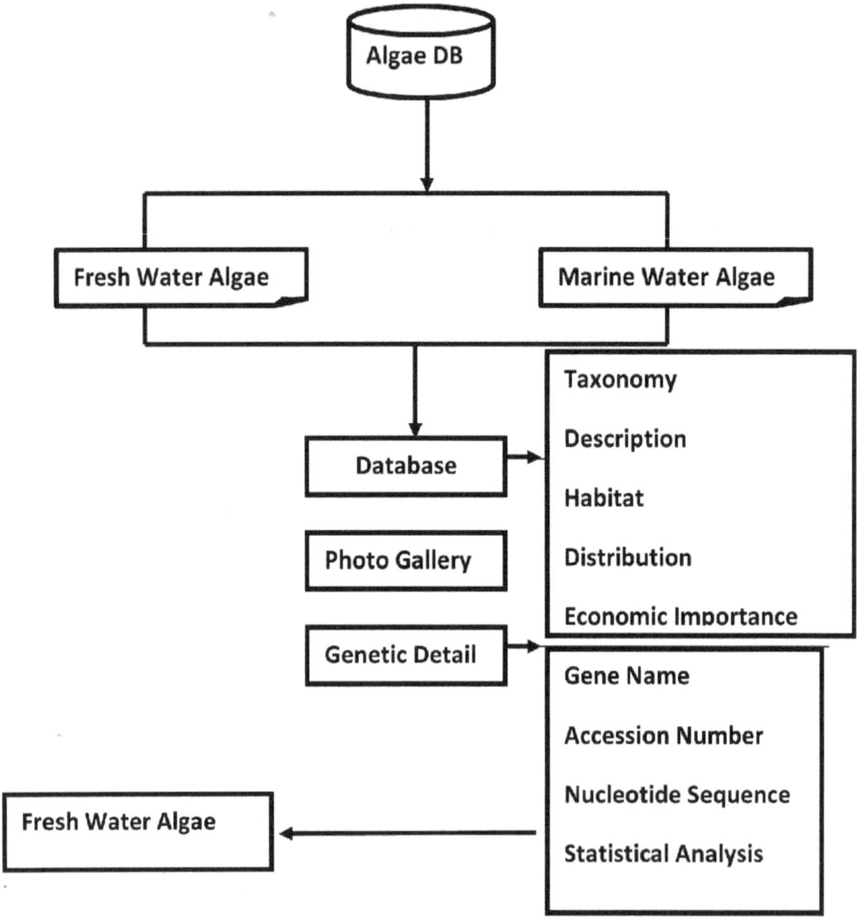

Fig1: The schema of the Database of Algae Resources in Andaman and Nicobar Islands

ALGAE NUCLEOTIDE INFORMATION

The database includes a collection of around 110 nucleotide sequences present in different species algae. It contains complete and partial genomic sequences, which includes genomic DNAs, chloroplast DNA, messenger RNA and ribosomal RNA. This database allows the identifiers to search the nucleotide sequence information in different field via. Accession number, Species name, Gene name etc. The nucleotide sequences for this database have been taken from National Centre for Biotechnology Information (NCBI).

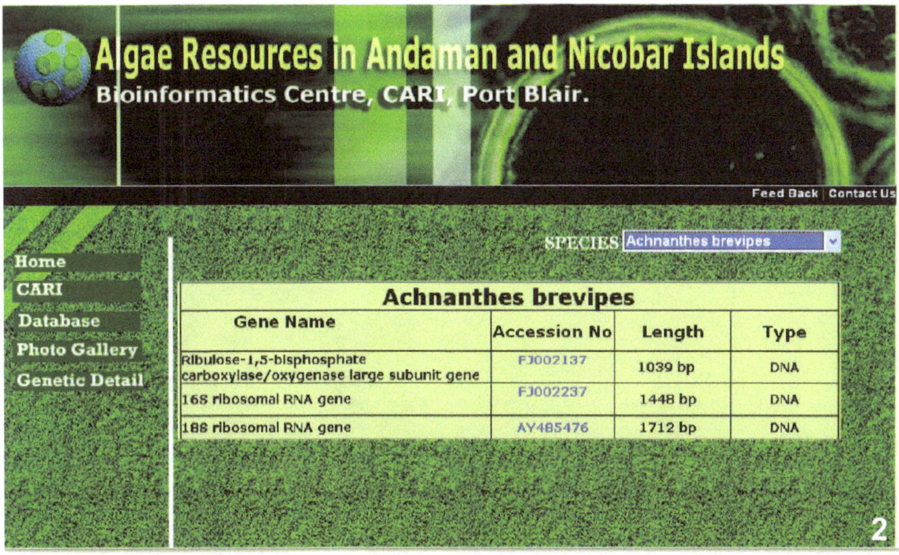

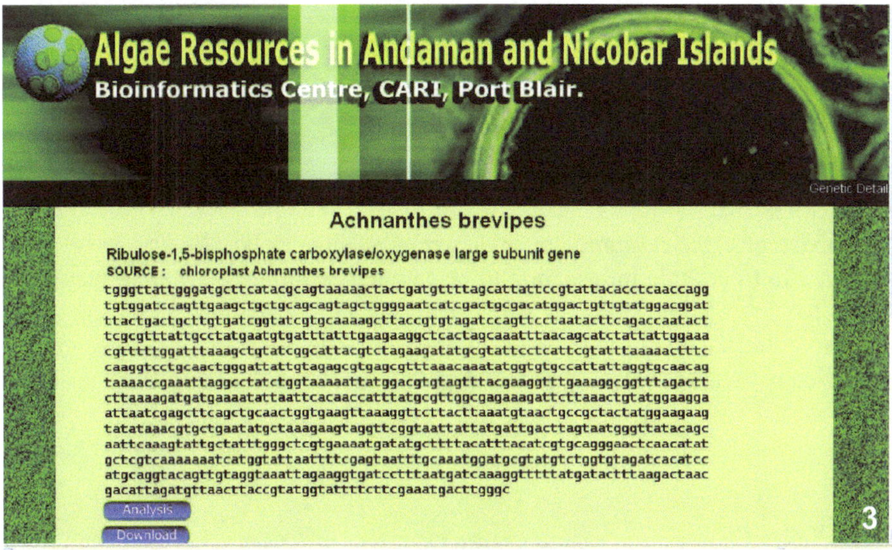

Fig 2,3: Retrieval of sequences using the drop down menu from database Statistical Analysis

DNA sequence with high percent of A+T nucleotides usually indicates intergenic parts of the sequence, while low A +T and high G+C nucleotide percentages indicates possible genes. Thus nucleotide sequences taken from National Centre for Biotechnology Information (NCBI) were analyzed using Bioedit software. This software helps to analyze the nucleotide composition of the sequence and provides the user information about the length of Nucleotide sequence, molecular weight of its single strand, double strand, its Guanine + Cytosine content, Adenine + Thiamine content, individual Mol percentage of Adenine, Guanine, Cytosine and Thiamine along with an graphical representation.

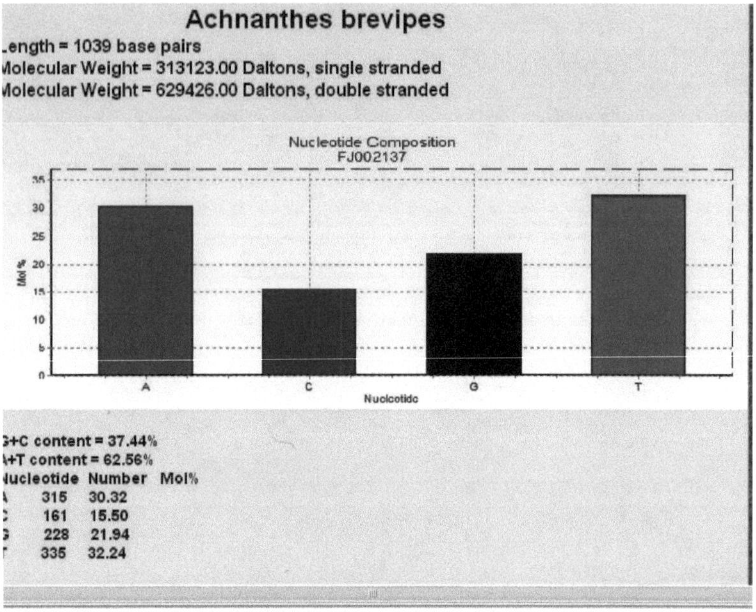

Fig. 4: Statistical Result

SUMMARY

A database of algae resources in Andaman and Nicobar Islands was developed with an aim to provide information to researchers working at different places in field related to marine and fresh water algal species. Algae photos were captured and also collected from secondary sources from different parts of the Island and integrated with other morphological and genetic information of respective species.

18

Bioinformatics of Natural Antibiotics

S. Rajarajan, M. Yogambal, K. Sangeetha

ANTIBIOTICS

Antibiotic is a drug used to treat infections caused by bacteria and other microorganisms. Technically an antibiotic is a substance produced by a microorganism that selectively inhibits the growth of another. Penicillin was the first antibiotic discovered by Alexander Fleming in 1928, a significant breakthrough for medical science. Antibiotics had saved thousands of wounded soldiers in the I and II world wars. Also, use of it made surgery safer and increased the average life expectancy of human beings. Antibiotics are among the most frequently prescribed medications in modern medicine. Some antibiotics are 'bactericidal', meaning that they work by killing bacteria. Each type of antibiotic affects different bacteria in a different way. Some antibiotics can be used to treat a wide range of infections and are known as 'broad-spectrum' antibiotics. Others are only effective against a few types of bacteria and are called 'narrow-spectrum' antibiotics.

Yellapragada Subbarao (January 12, 1895-August 9, 1948) is a name to be remembered by all Indians. He was the only Indian to discover an antibiotic namely Aureomycin in 1945, a semi synthetic antibiotic under category of tetracycline. Antibiotics are produced by a number of microorganisms and inhibit the growth of other microorganisms even at very low concentrations. The antibiotics have wide application in chemotherapy, plant pathology, food preservation, veterinary medicine and as research tools in biochemistry and molecular biology.

Natural antibiotics used to treat infection includes Amphotericin, Cephalosporin, Chloramphenicol, Erythromycin, Gentamycin ,Kanamycin ,Neomycin ,Penicillin,

Streptomycin, Vancomycin, Tetracycline, Lincomycin, Bacitracin, Polymyxin, Thienamycin, Clavulanicacid, Lincomycin, Theinamycin, Tetracycline, Monensin, Tobramycin, Cephamycin, Fusidic acid, Puromycin and Gramicidin. The genomic information of these antibiotic secreting regions in bacteria is very relevant for discovery or development of newer antibiotics.

APPLICATION OF BIOINFORMATICS ON GENE ANALYSIS

Deciphering of complete genome sequences of bacterial organisms has revolutionized the search for new antibiotics. The completion of about 30 bacterial whole-genome sequences and ongoing sequencing projects of over 100 microbial organisms will allow researchers to find novel therapeutic targets in innovative ways.

The difficulty is the massive comprehensive and application of data obtained from the genome sequences. The sequence of microbial pathogens catalogs every gene product that would be relevant for the host-parasite interaction and potential antibiotic drug target.

Therefore, scientists interested in discovering antibiotics should utilize theinformation from genomes through comparative, functional, or structural genomics in order to simplify drug target selection. The present review on the subject reveals that out of the known 24 natural antibiotics, genomic information of the microorganism pertaining to antibiotic production is available for 17 while for the other seven antibiotics, genomic information has to be explored.

AMPHOTERICIN B

Amphotericin B is produced by Septromyces nodusus and it was first discovered by Dr.Selman Abraham Waksman in1952. Amphotericin B can be used for topical applications, as well as, through intravenous for the treatment of mycotic infections such as Blastomycosis, Systemic candidiasis, Streptococcosis, Moniliasis etc. Amphotericin acts by destroying the integrity of the cellular membrane of susceptible organisms. This action is related to binding of the polyenes to steroids in the membrane and the formation of pores. This mechanism explains the absence of antibacterial activity since bacterial membrane lack steroids. The adverse side reactions upon systemic administration of Amphotericin include haemolytic anemia due to alteration in erythrocyte membranes. The most frequently observed toxicity with systemic use of Amphotericin B is nephrotoxixity , which is usually reversible upon cessation of therapy .A low concentration of amphotericin B in combination with other selected antibiotics such as rifampicin and tetracycline seem to facilitate membrane passage of the latter antibiotics, which normally lack the ability to penetrate into the fungal cell.

SUSCEPTIBLE ORGANISM

Aspergillus spp, *Blastomyces* spp , *Candida* spp.(execept *Candida krusei* and

Candida lusitania), *Coccidioides* spp.,*Cryptococcus* spp.,*Histoplasma* spp., are susceptible to Amphotericin B.

GENOMIC INFORMATION

Amphotericin gene cluster was found to include amphRI, RII, RIII and RIV genes that are similarto regulatory genes in other polyene biosynthetic gene clusters. Regulatory genes, amphRI, wasfound to have a homologue, amphRVI, located in the right flank at a distance of 127 kbp along the chromosome. *Aspergillus* spp., *Blastomyces* spp., *Candida* spp. (execept *Candida krusei* and *Candida lusitania*), *Coccidioides* spp.,*Cryptococcus* spp., *Histoplasma* spp., are susceptible to Amphotericin B.

GENOMIC INFORMATION

Amphotericin gene cluster was found to include amphRI, RII, RIII and RIV genes that are similar to regulatory genes in other polyene biosynthetic gene clusters .Regulatory genes, amphRI, wasfound to have a homologue, amphRVI, located in the right flank at a distance of 127 kbp along the chromosome.

CEPHALOSPORIN

In 1945 Broutz isolated a microorganism from seawater, which was collected near sewage outlet of the coast of Sardinia. This organism was identified as cephalosporium acremonium. Abraham andco-workers at Oxford reported the isolation of three substances with antibiotic activity fromculture of this organism during 1955-56. These metabolites were Cephalosporin P (Steroid),penicillin N and Cephalosporin C. The degree of antibacterialactivity of cephalosporin C is only moderate and is not used therapeutically. However on hydrolysis it yields cephalosporinic acid, which serves as starting material for the semisynthetic production of large number of cephalosporin antibiotics. It was not possible to obtain 7-aminicephalosporinic acid (7-ACA) either by fermentation or by enzymatic hydrolysis of cephalosporin C. Therefore 7-ACA is obtained from cephalosporin C by chemical hydrolysis.

GENOMIC INFORMATION

Cephalosporin is produced by condensation of the three precursor aminoacids L-á-aminoadipic acid,L-cysteine and D –valine to form the tripeptibe ä-L-á-aminoadipyl-L-cysteinyl-D-Valine.This reaction is carried out by ACV synthetase which is encoded by an unusually large gene (pcbAB).The tripeptide ACV is cyclised to form isopenicillin N(ipn) by the enzyme IPN synthase encoded by the pcbC gene and the IPN is isomerised to the D-configuration, giving rise to Penicillin N,which is later converted to deacetoxycephalosporin C by a ring expansion step carried out by the deacetoxycephalosporin C synthase encoded by the cefEF gene. The latter enzyme is also able to oxidise the methyl group at carbon 3 of deacetooxycephalosporin C to give deacetylcephaloaporin (DAC). Both activities are located in a single polypeptide that is encoded by the cefEF gene.The last step of the cephalosporin biosynthetic pathway involves the conversion of DAC to cephalosporin C by the enzyme acetyl coenzyme A : DAC acetyltransferase. (Santiago Gutierrez *et al.,* 1992).

ERYTHROMYCIN

Erythromycin is produced from *Streptomyces erythreus* by Dr.Selman Abraham Waksman, 1959. Erythromycin A has a fourteen membered macrocyclic lactone ring having 2-sugar residues namely desosamine and cladinose attached at positions C-5 and C-5 through glycosidic bond. Erythromycin has a large number of asymmetric centers. The macrolactone ring is unstable in gastric acidity. Therefore enteric coating can be used to deliver the antibiotic to the intestinal tract where it is readily absorbed. Erythromycin exerts its action by inhibiting the protein synthesis. It binds to the 50S subunit of microbial 70S ribosomes and also blocks translation of ribosomes. Erythromycin is used primarily in treating Pneumococcal and haemolytic streptococcal infections.

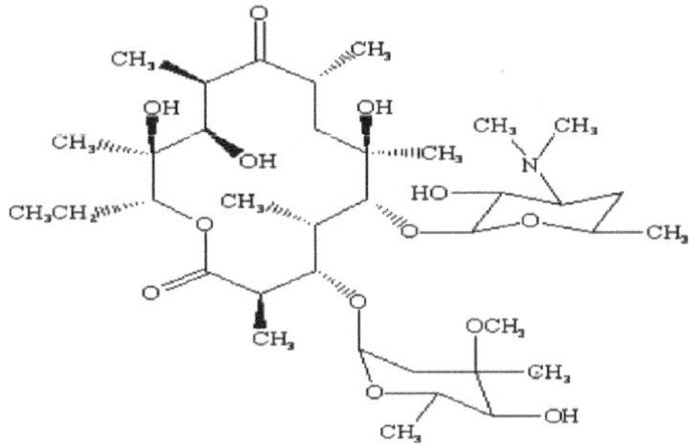

Fig: Erythromycin

GENOMIC INFORMATION :

Biosynthetic genes are clustered on the *Streptomyces erythraeus* chromosome similarly to other secondary metabolic pathway genes.

- The erythromycin gene cluster contains 20 genes involved in the biosynthesis of erythromycin.
- Production requires propionyl-CoA to provide a starter unit, and (2s)-methylmalonyl-CoA to provide extender units for the polyketide chain of the antibiotic..

KANAMYCIN

Kanamycin is synthesised form *Streptomyces kanamyceticus* by Dr. Hamao Umezawa,1957. Kanamycin sulfate is an aminoglycosideantibiotic, available in

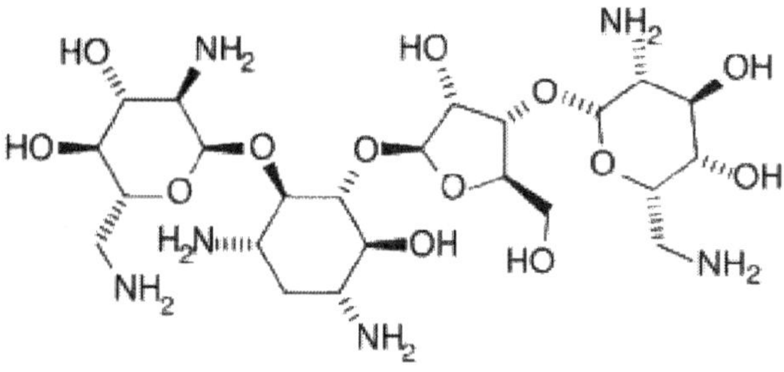

Fig: Kanamycin

both oral and intravenous forms, and used to treat a wide variety of infections .Kanamycin works by affecting the 30S ribosomal subunit and causes a frameshift mutation or it prevents the translation of RNA. This means that instead of a codon CAT (for example in sequence CATG), a codon ATG is read by aminoacyl tRNA (aa-tRNA). Aminoacyl tRNA is consequently carrying a different amino acid, because the anticodon on the aa-tRNA is different.

Staphylococcus epidermidis, Neisseria gonorrhoeae, Haemophilus influenza, Escherichia coli, Enterobacter aerogenes, Shigella sp., Salmonella species, Klebsiella pneumoniae., Serratia marcescens, Providencia species, Acinetobacter species, Citrobacter freundii, Citrobacter species are susceptible to Kanamycin.

GENOMIC INFORMATION

The sequencing of the 47 kb-region of *S. kanamyceticus* genomic DNA revealed 40 putative open reading frames (ORFs) encoding Km biosynthetic proteins, regulatory proteins, and resistance and transport proteins(Madan K. Kharel *et al.,*2004).

NEOMYCIN

Commercial Neomycin is a complex mixture of aminoglycoside antibiotics originally isolated from a culture of *Streptomyces fradiae* by Waksman *et al* in 1949. The principle components of the mixture are Neomycin B and Neomycin C together with a small quantity of neamine. Neomycin is an aminoglycosideantibiotic that is found in many topical medications such as creams, ointments and eyedrops. Neomycin is overwhelmingly used as a topical preparation, such as Neosporin. It can also be given orally, where it is usually combined with other antibiotics. Neomycin is not absorbed from the gastrointestinal tract, and has been used as a preventive measure for hepatic encephalopathy and hypercholesterolemia. By killing bacteria in the intestinal tract, it keeps ammonia levels low and prevents hepatic encephalopathy, especially prior to GI surgery.

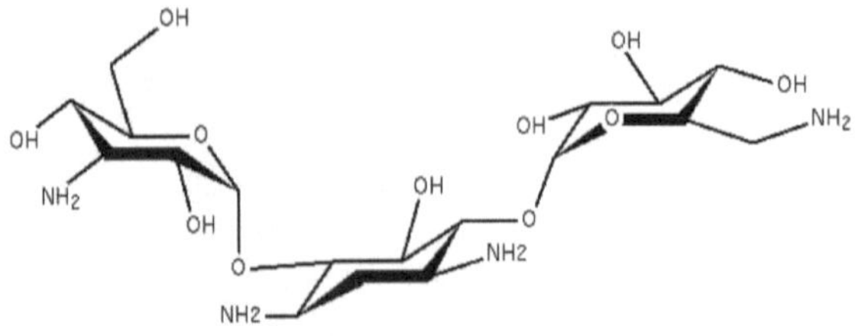

Fig: Neomycin

Gene	Details
Topology	Linear
Length (bp)	8,545,929
G+C Content (%)	72.2
Protein-Coding Gene	7,138
rrn Operon (16S-23S-5S)	6
tRNA Gene	66

GENOMIC INFORMATION

Sequence analysis of *Streptomyces fradiae* NCIMB 8233 has identified 21 putative open reading frames (ORFs) in the neomycin gene cluster (*neo*) with significant protein sequence similarity to gene products involved in the biosynthesis of other DOS-containing aminoglycosides, namely butirosin (*btr*), gentamycin (*gnt*), tobramycin (*tbm*) and kanamycin (*kan*). Located at the 5-end of the *neo* gene cluster is the previously-characterised neomycin phosphotransferase gene (*apH*). Three genes unique to the *neo* and *btr* clusters have been revealed by comparison of the *neo* cluster to *btr*, *gnt*, *tbm* and *kan* clusters. This suggests that these three genes may be involved in the transfer of a ribose moiety to the DOS ring during the antibiotic biosynthesis.(Fanglu Huang *et al.,* 1995).

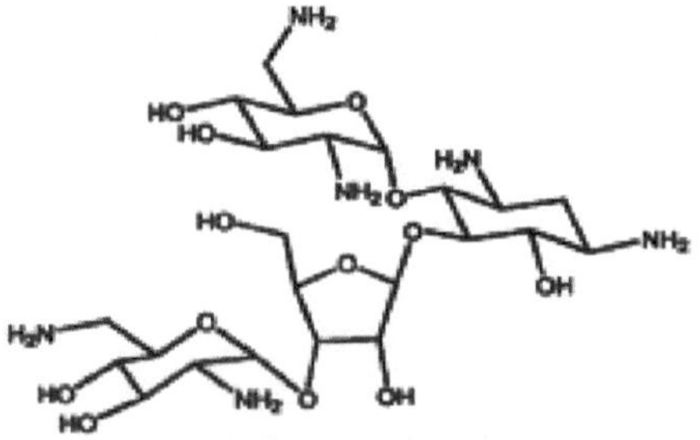

Neomycin Biosynthetic Gene Cluster in *Streptomyces fradiae*

PENCILIN

Penicillin (sometimes abbreviated PCN or pen) is a group of antibiotics derived from *Penicilliumspecies* by Alexander Fleming,1928.Penicillin antibiotics are historically significant because they are the first drugs that were effective against many previously serious diseases such as syphilis and Staphylococcus infections. Penicillins are still widely used today, though many types of bacteria are now resistant. All penicillins are Beta-lactam antibiotics and are used in the treatment of bacterialinfections caused by susceptible, usually Gram-positive, organisms.

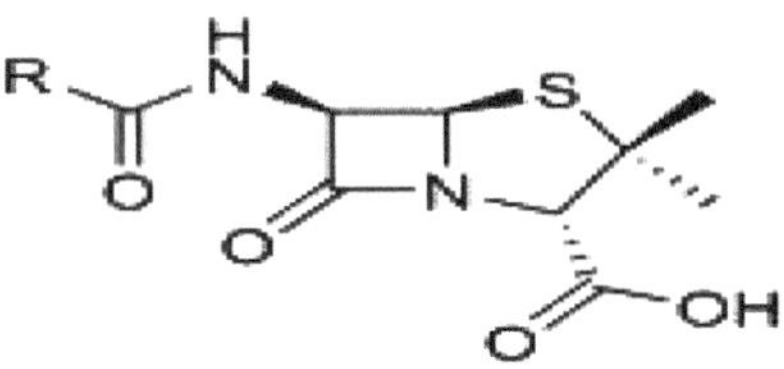

Fig : Pencilin

GENOMIC INFORMATION

The genes pcbAB, pcbC and penDE encodes the enzymes (alpha-aminoadipyl-cysteinyl-valine synthetase, isopenicillin N synthase and isopenicillin N acyltransferase, respectively) involved in the biosynthesis of penicillin. Each gene is expressed as a single transcript from separate promoters. It is mediated by the zinc-finger transcription factor PacC, an activator of transcription of the isopenicillin N synthase gene.

STREPTOMYCIN

Streptomycin is synthesized from *Streptomyces griseus* by Dr.Selman Abraham Waksman, 1944.It is the first of a class of drugs called aminoglycosides to be

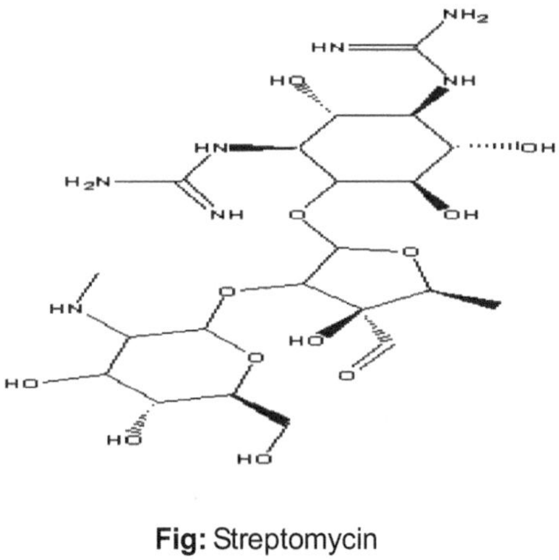

Fig: Streptomycin

discovered, and was the first antibiotic remedy for tuberculosis. Streptomycin is a bactericidal antibiotic. Streptomycin cannot be given orally, but must be administered by regular intramuscular injections. An adverse effect of this medicine is ototoxicity, which can lead to hearing loss. The importance in pharmacogenomics (genetic based variation is response to drug)indicate careful analyses of individual variation response to this drugs.

GENOMIC INFORMATION

A-factor is required for both secondary metabolism (streptomycin and grixazone Production) and morphological differentiation. A-factor requires afsA gene for its synthesis, in regulating the streptomycin production. A-factor binds to its cytoplasmic binding protein ArpA and releases the adpA promoter and allowing adpA transcription. AdpA is required for activation of transcription of strR. A pathway-specific regulatory gene for streptomycin production and adpA regulon, that are required for morphological differentiation. adpA is the only ArpA-dependent gene.

NYSTATIN

Nystatin is synthesied from *Streptomyces noursei* by Rachael F. Brown and Elizabeth Hazen,1970.Nystatin is a polyeneantifungaldrug to which many molds and yeast infections are sensitive, including *Candida* spp. Nystatin has some toxicity associated with it when given intravenously, but it is not absorbed across intact skin or mucous membranes. It is considered a relatively safe drug for treating oral or gastrointestinal fungal infections.

GENOMIC INFORMATION

Six putative regulatory genes are located at the flank of the nystatin biosynthetic gene cluster. Gene inactivation and complementation experiments revealed that nysRI, nysRII, nysRIII, and nysRIV are necessary for efficient nystatin production.nificant roles could be demonstrated for the other.

BACITRACIN

Bacitracin is a mixture of related cyclic polypeptides produced by organisms of the licheniformis group of *Bacillus subtilisvar* Tracy, isolation of which was first reported in 1945.As it is a toxic and difficult-to-use antibiotic, bacitracin doesn't work well orally. However, it is very effective topically. Its action is on gram positive cell walls. Bacitracin shows susceptibility to *Staphylococcus* and *Micrococcus* Species

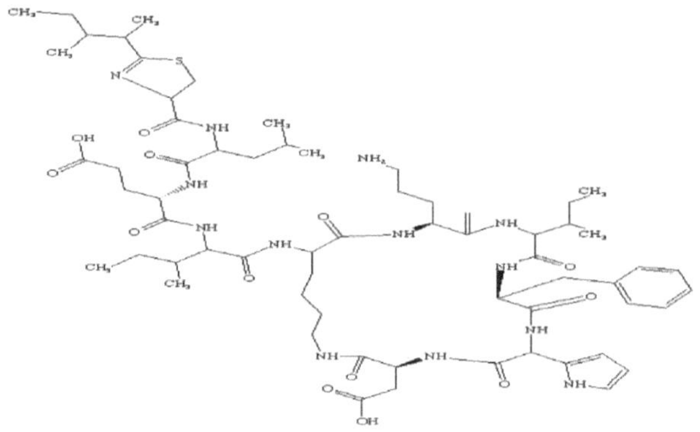

Fig : Bacitracin

CHLORAMPHENICOL

Chloramphenicol is a broad – spectrum antibiotics, which was first isolated from *streptomyces venezuelaes* in 1947. Subsequently it was identified and synthesized. Chloromycetin was the first antibiotic produced synthetically on a commercial scale. Chemically Chloramphenicol is a simple phenylpropanoid. It gained considerable attention, as it was first truly broad- spectrum antibiotic to be discovered. Its spectrum of action includes gram-positive bacteria, a few virus and a number of rickettsial pathogens.

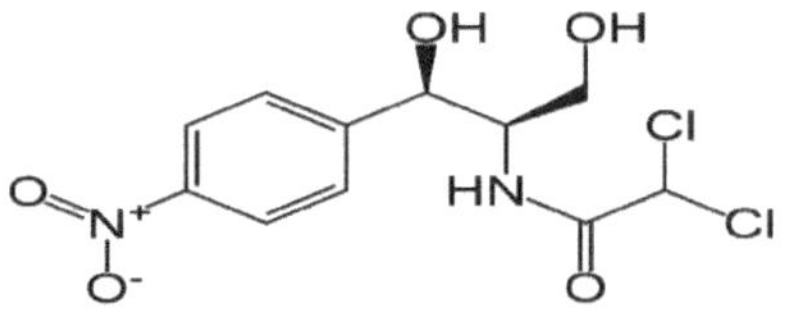

Fig: Chloramphenicol

Neisseria meningitidis, Streptococcus pneumoniae, Haemophilus influenza, Enterococcus faecium, Salmonella typhi, Vibrio cholera, Brucella species, Bordetella pertussis, Rickettsiae spp., Mycoplasma spp., are susceptible organism to chloramphenicol .

GENOMIC INFORMATION

pabAB, an amino-deoxychorismate synthase gene is responsible for chloramphenicol(Cm) production.(J. He.,2001)

VANCOMYCIN

Vancomycin was first isolated by EC Kornfeld (working at Eli Lilly) from a soil sample collected from the interior jungles of Borneo by a missionary. The organism that produced it was eventually named *Amycolatopsis orientalis* .The original indication for vancomycin was for the treatment of penicillin-resistant *Staphylococcus aureus*.

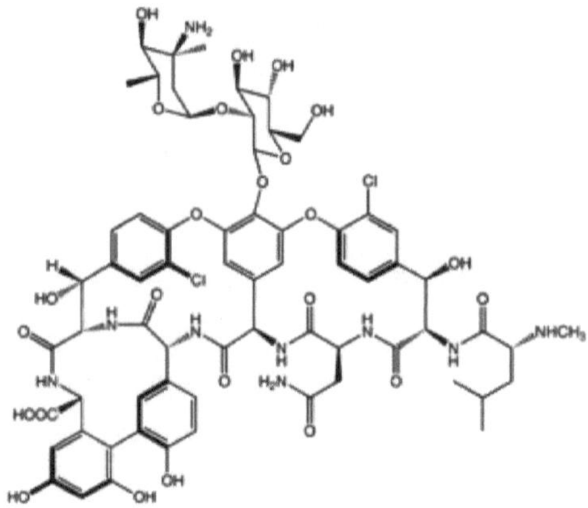

Fig : Vancomycin

GENTAMYCIN

Gentamycin is synthesied from *Micromonospora purpureochromogenes* by Dr. Marvin Weinstein, 1963. Gentamycin C1 is composed of three sugar moieties consisting of purpurosamine A, Garosamine attached to 2-deoxy–streptamine through glycoside bonds. Gentamycin is rapidly absorbed on intramuscular administration and it is readily distributed in various body parts. The peak serum levels are often achieved in less than one hour. Gentamycin is available in formulations for topical use, but its principle use is parenternal for the treatment

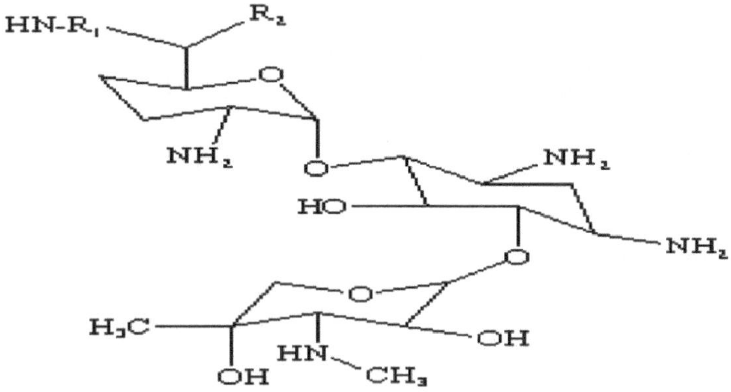

Fig: Gentamycin

of serious Gram-negative infections caused by sensitive organisms (e.g. *Enterobacter, Echerichia & Klebsilla*). The combination of carbenicillin with Gentamycin is available as sulphate salt and the usual adult dose is the equivalent of 1 mg Gentamycin per kg of body weight intramuscularly or by intravenous injections.

Enterobacter species, Streptococcusfaecalis,Escherichia coli, Klebsiellaspecies, Proteus mirabilis,Proteus species,Pseudomonas aeruginosa,Serratia species are susceptible organism to Gentamycin

PLYMIXIN

Polymyxin B (PMB), a cyclic cationic peptide obtained from *Bacillus polymyxa* was discovered in 1947. It has been used considerably for treating certain bacterial infections such as the meningial infections caused by *Haemophilus influenzae* , urinary tract infections of *Escherichia coli* and bacteremia caused by *Enterobacter aerogenes* etc.(Vikrant M *et al.*,2005)

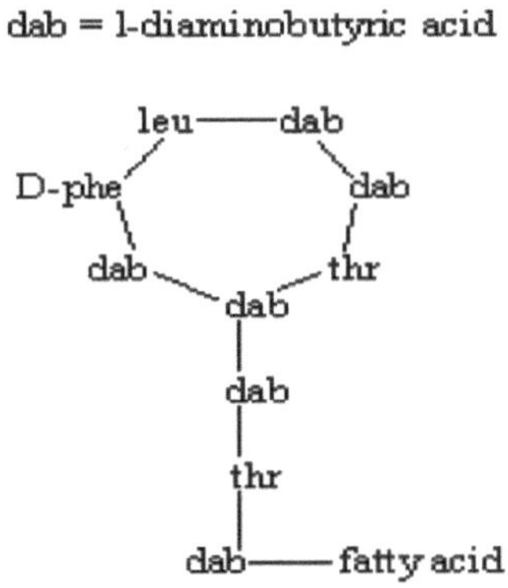

Fig: Plymixin

CLAVULANIC ACID

Clavulanic acid (rINN) is a beta-lactamase inhibitor sometimes combined with penicillin group of antibiotics to overcome certain types of antibiotic resistance. Specifically, it is used to overcome resistance in bacteria that secrete beta-lactamase enzymes, which otherwise inactivate most penicillins. Most commonly, the potassium salt potassium clavulanate is combined with amoxicillin (co-amoxiclav) or ticarcillin. The name is derived from the *Streptomyces clavuligerus* microorganisms from which clavulanic acid is derived. Clavulanic acid is biosynthetically generated from the amino acid arginine.(Coryell MW *et al.*,2009)

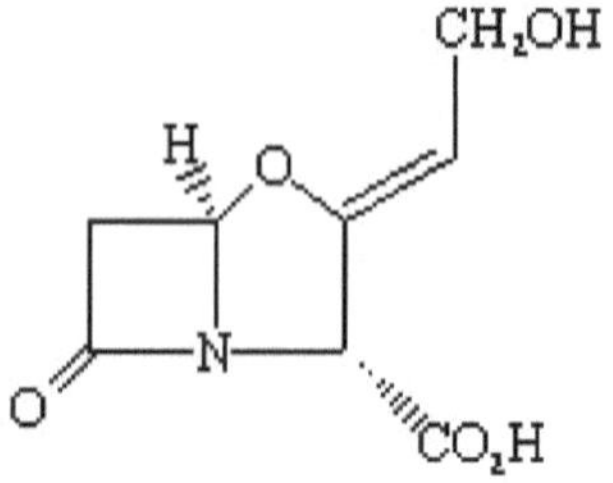

Fig : Clavulanic Acid

GENOMIC INFORMATION

Jensen *et al.*, characterized an approximately 15-kbstretch of chromosomal DNA located adjacent to the cephamycin gene cluster in *S. clavuligerus* and demonstrated that the openreading frames (ORFs) were involved in clavulanic acid production. Genes located in this region include *ceaS* (encoding carboxyethylargininesynthase), *bls* (β-lactam synthetase), *pah* (proclavaminate amidinohydrolase), *cas2* (clavaminatesynthase) and *cad* (clavulanic acid dehydrogenase), all enzymes are required in the biosynthetic pathway.In addition, *oat*, a gene encoding a protein with ornithine acetyltransferaseactivity, has been identified, but its role in clavulanicacid is not yet understood, and *claR*, a gene encoding a transcriptional regulator that controls the late steps in clavulanic acid biosynthesis,is also located in this region . Two additional genes, *orf7* and *orf10*, encode an apparent peptide transport protein and a cytochrome P450 type protein and are essential for clavulanic acid biosynthesis. (S.E. Jensen *et al.*, 2004).

LINOMYCIN

Lincomycin was isolated in 1962 from soil acinomycete found near Lincoln, Nebraska, which gave origin to its name (Stratton 1998). The actinomycete was named as *Streptomyces lincolnesis var, Lincolnensis, a new streptomycete* species. Lincomycin proved to be a member of a new class of antibiotics characterised by a alkyl 6 amino 6s,8-dideoxy – 1-D-erythro – alpha-D-alacto-octopyranoside joined with a proline moiety by an amide linkage. (J. Spiker and T. Renzaka., 2004)

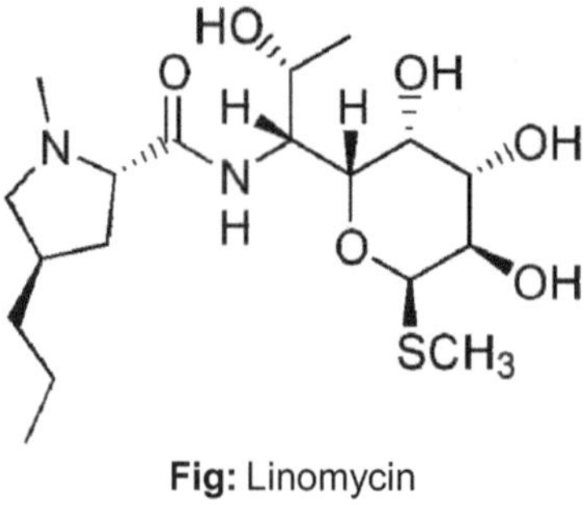

Fig: Linomycin

GENOMIC INFORMATION

The *lmb/lmr* gene cluster is composed of 27 open reading frames with putative biosynthetic or regulatory functions (*lmb* genes) and three resistance *(lmr)* genes, two of which, *lmrA* and *lmrC*, flank the cluster.

THIENAMYCIN

Thienamycin, â-lactam antibiotic with the unique structure was discovered in the course of screening soil microorganisms for the production of inhibitors of peptidoglycan synthesis in Gram-positive and Gram-negative bacteria. Taxonomic studies of the producing organism MA4297 resulted in its assignment to a new *streptomycete* species which has been named *Streptomyces cattleya*. (J.S. Kahan *et al.,* 1978)

GENOMIC INFORMATION

Two regulatory genes, *thnI* and *thnU* are involved in the biosynthesis of Thienamycin. Gene expression analysis applied to the *thn* gene cluster demonstrated that ThnI is a transcriptional activator essential for thienamycin biosynthesis that regulates the expression of nine genes involved in thienamycin assembly and export (*thnH*, *thnJ*, *thnK*, *thnL*, *thnM*, *thnN*, *thnO*, *thnP* and *thnQ*). (Miriam Rodríguez *et al.*, 2008).

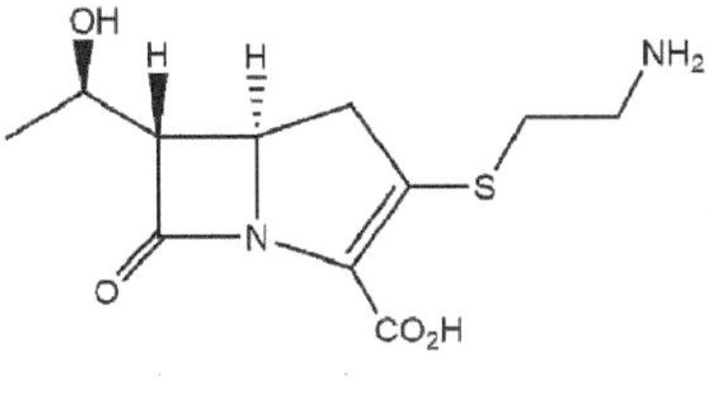

Fig: Thienamycin

NOVOBIOCIN

Novobiocin, also known as albamycin or cathomycin, is an aminocoumarin antibiotic that is produced by the actinomycete *Streptomyces niveus*. Novobiocin was first reported in the mid-1950s (then called streptonivicin).

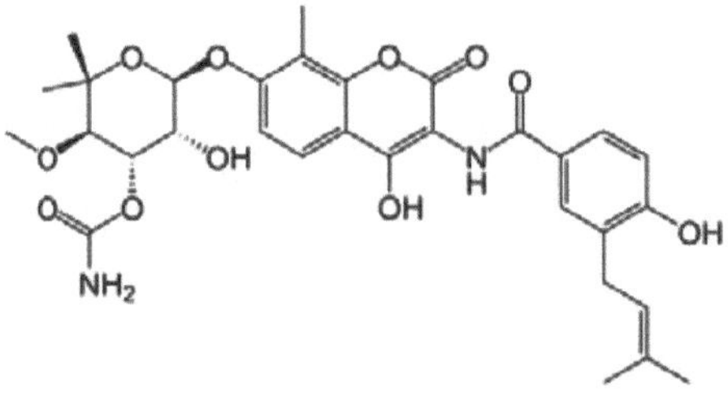

Fig: Novobiocin

TETRACYCLINE

Tetracycline in synthesised from *Streptomyces aureofaciens* by Lloyd Conover, 1948. Tetracycline is a broad-spectrum polyketideantibiotic produced by the *Streptomyces*genus of Actinobacteria, indicated for use against many bacterial infections. It is a protein synthesis inhibitor. It is commonly used to treat acne today, and more recently, rosacea, and played a historical role in stamping out cholera in the developed world.

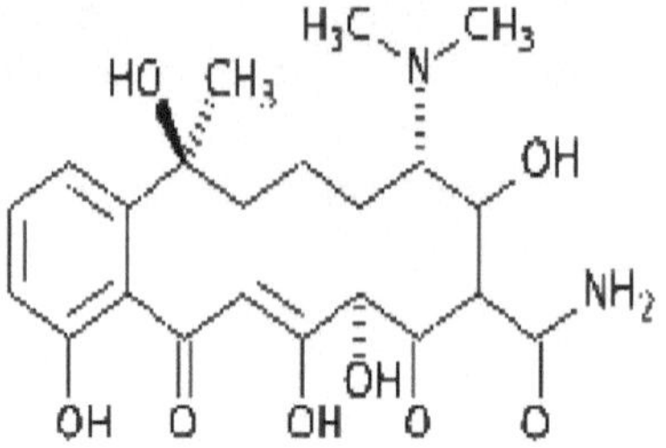

Fig : Tetracycline

MONENSIN

Monensin belong to class of polyether antibiotics.It is produced predominantly by *Streptomyces cinnamonensis.* It is highly effective against Staphylococci and are widely used as coccidiostats. (S.V. Bhat *et al.,* 2008). The structure of monensin was first described by Agtarap *et al.,* in 1967, and was the first polyether antibiotic to have its structure elucidated.

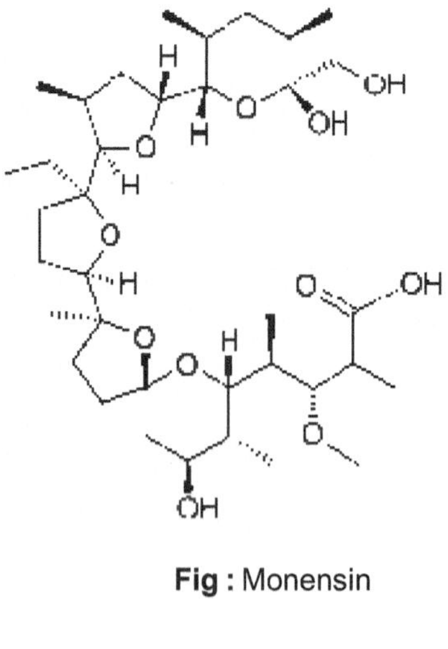

Fig : Monensin

GENOMIC INFORMATION

A candidate biosynthetic gene cluster (97 kbp) for the polyether ionophore monensin from Streptomyces cinnamonensis has revealed a modular polyketide synthase composed of eight separate multienzyme subunits housing a total of 12 extension modules, and flanked by numerous other genes for which a plausible function in monensin biosynthesis has to be ascribed. (Oliynyk M*et al.*,2003).

TOBRAMYCIN

Tobramycin is structurally related to Kanamycin B differing only in the absence of 3- hydroxyl group in the Kanosamine residue and the presence of amino group. It was obtained from *S. tenebrarius* in 1957 by Lilly research Laboratories. It is used to treat various types of bacterial infections, particularly Gram-negative infections.Tobramycin works by binding to a site on the bacterial 30S and 50Sribosome, preventing formation of the 70S complex. As a result, mRNA cannot be translated into protein and cell death ensues.

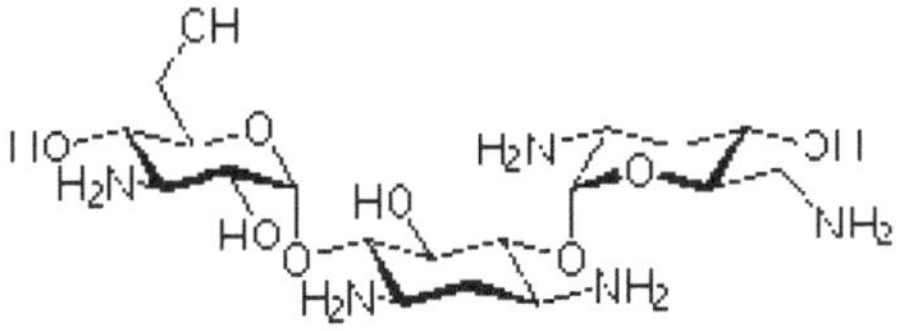

Fig : Tobramycin

GENOMIC INFORMATION

Sequencing of 33.9 kb has revealed 24 open reading frames (ORFs) including putative tobramycin biosynthetic genes

CEPHAMYCIN

Cephamycins are a group of beta-lactam antibiotics. They are very similar to cephalosporins, and the cephamycins are sometimes classified as cephalosporins. Like cephalosporins, cephamycins are based upon the cephem nucleus. Cephamycins were originally produced by *Streptomyces*. Cephamycins possess a methoxy group at the 7-alpha position.

GENOMIC INFORMATION

Three open reading frames (ORFs) have been located downstream of *cefE* in the cephamycin C gene cluster of *Streptomyces clavuligerus*. ORF13 (*pcd*) encodes a 496-amino-acid protein (molecular weight [MW], 52,488) with an N-terminal amino acid sequence identical to that of pure piperideine-6-carboxylate dehydrogenase. ORF14 (*cmcT*) encodes a 523-amino-acid protein (MW, 54,232) analogous to *Streptomyces* proteins for efflux and resistance to antibiotics. ORF15 (*pbp74*) encodes a high molecular weight penicillin-binding protein (MW, 74,094, Francisco J.,1998).

FUSIDIC ACID

Fusidic acid was isolated from the fermentation products of the fungus Fusidium coccineum. Fusidic acid shows a high antibiotic activity against staphylococcus. It is an antibiotic which contains the cyclopentanoperhydrophenanthrene ring system and thus belongs to the steroids or tetracyclic triterpenes. (Victor Wynn.,1965). Fusidic acid, a fusidane that interferes with protein synthesisvia the translocase enzyme, is mainly notable for its activityagainst staphylococci, coagulase-positive and negative, whetheror not they are resistant to methicillin and related penicillins. It is also active against coryne bacteria and against many genera of strict anaerobes and microaerophiles. (L.Verbist *et al.,* 1990).

PUROMYCIN

Puromycin is an aminonucleosideantibiotic, derived from the Streptomyces alboniger bacterium, that causes premature chain termination during translation taking place in the ribosome. Resistance to puromycin is conferred by the Pac gene encoding a puromycin N-acetyl-transferase (PAC) that was found in a Streptomyces producer strain.

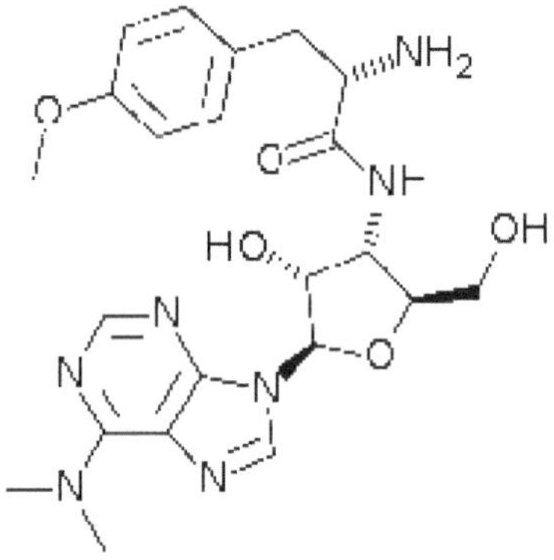

Fig : Puromycin

GENOMIC INFORMATION

Puromycin biosynthesis (Pur cluster) was encoded by a single DNA fragment of 15 kb.

GRAMICIDIN

Gramicidin is a heterogeneous mixture of six antibiotic compounds, Gramicidins A, B and C, making up 80%, 6%, and 14% respectively,[1] all of which are obtained from the soil bacterial species *Bacillus brevis* and called collectively Gramicidin D. Gramicidin D are linear pentadecapeptides; that is chains made up of 15 amino acids.[2] This is in contrast to Gramicidin S which is a cyclic peptide chain. Gramicidin is active against Gram-positive bacteria, except for the Gram-positive bacilli, and against select Gram-negative organisms, such as Neisseria bacteria.

Its therapeutic use is limited to topical application as it induces hemolysis in lower concentrations than bacteria cell death, thus cannot be administered internally. (Bourinbaiar, A.S., and C.F. Coleman. 1997).

REFERENCES

Bhat, S.V., Nagasampagi, B.A., Sivakumar. M., Chemistry of Natural products.Narosa publishing house.749-750.

Bourinbaiar, A.S., and C.F. Coleman. 1997. The effect of gramicidin, *Arch. Virol.* 142: 2225-2235.

Clavulanic acid: A competitive inhibitor of beta-lactamases with novel anxiolytic-like activity and minimal side effects.

Coryell, M.W., Wunsch, A.M., Haenfler, J.M., Allen, J.E., Schnizler, M., Ziemann, A.E., Kim, D.J., King, J.A., Zuccarelli, L., Ferris, C.F., Koppel, G.A., Snowdon, C.T. and Ahn, C.H. 2009. *Pharmacology Biochemistry and Behavior*, 93(2):112-120.

Francisco, J. Pérez-Llarena, Antonio Rodríguez-García, Francisco J. Enguita, Juan, F. Martín, and Paloma Liras. 1998. The pcd Gene Encoding Piperideine-6-Carboxylate Dehydrogenase Involved in Biosynthesis of α-Aminoadipic Acid Is Located in the Cephamycin Cluster of Streptomyces clavuligerus. *J. Bacteriol.*, 180(17): 4753–4756.

He, J., Magarvey, N., Piraee, M. and Vining, L.C. 2001. The gene cluster for chloramphenicol biosynthesis in Streptomyces venezuelae ISP5230 includes novel shikimate pathway homologues and a monomodularnon-ribosomal peptide synthetase gene. *Genetics and Molecular biology*, 147:2817-2829.

Kharel, K., Bimala Subba, Devi B. Basnet, Jin Suk Woo, Hei Chan Lee, Kwangkyoung Liou and Jae Kyung Sohng. 2004. A gene cluster for biosynthesis of kanamycin from Streptomyces kanamyceticus: comparison with gentamicin biosynthetic gene cluster. *Archives of Biochemistry and Biophysics*, 429(2):204-214.

Kahan, J. S., Kahan, F. M., Goegelman, R., Currie, S. A., Jackson, M., Stapley, O., Miller, T. W., A. K. Miller, D.Hendlin, S. Mochalest, S. Hernandezt, H.B. Woodruff and J. Thienamycin 1979. A new β-lactam antibiotic 1. Discovery, taxonomy, isolation and physical properties. Birnbaum. *The Journal of Antibiotics*, 32(1):1-12 .

Verbist, L. 1990. The antimicrobial activity of fusidic acid. *The British Society for Antimicrobial Chemotherapy, Journal of Antimicrobial Chemotherapy,* 25:1-5..

Lincomycin, 2004. Cultivation of producing stains and biosynthesis. J.Spiker, T.Renzaka. Applied microbiology and biotechnology.63 No5.

Miriam Rodríguez, Luz Elena Núñez, Alfredo F. Braña, Carmen Méndez, José A. Salas and Gloria Blanco* 2008. Identification of transcriptional activators for thienamycin and cephamycin C biosynthetic genes within the thienamycin gene cluster from Streptomyces cattleya. *Molecular Microbiology*, 69(3):633-645.

Oliynyk, M., Stark, C.B., Bhatt, A., Jones, M.A., Hughes-Thomas, Z.A., Wilkinson, C., Oliynyk, Z., Demydchuk, Y., Staunton, J., Leadlay, P.F. 2003. Analysis of the biosynthetic gene cluster for the polyether antibiotic monensin in Streptomyces cinnamonensis and evidence for the role of monB and monC genes in oxidative cyclization. *Mol. Microbiol.* 49(5):1179-90.

Polymyxin, B. 2005. An ode to an old antidote for endotoxic shock. Vikrant M. Bhor,a Celestine J. Thomas,{a Namita Suroliab and Avadhesha Surolia*a.Bhor, Vikrant M and Thomas, Celestine J and Surolia, Namita and Surolia, Avadhesha P. *In: Molecular BioSystems*, 1(3):213-222.

Jensen, S.E., Paradkar, A.S., Mosher, R.H., Anders, C., Beatty, P. H., Brumlik, M.J., Griffin, A. and Barton, B. 2004. Five Additional Genes Are Involved in Clavulanic Acid Biosynthesis in Streptomyces clavuligerus. *Antimicrobial Agents and Chemotherapy*, 48(1):192-202.

Santiago Gutierrez, Javier Velasco, Francisco, J. Fernandez and Juan F. Martin. 1992. The cef G gene of Cephalosporium acremonium is linked to the cefEF gene and encodes a Deacetylcephalosporin C acetyltransferase closely related to Homoserine O-acetyltransferase. *Journal of Bacteriology*, 3056-3064.

The neomycin biosynthetic gene cluster of Streptomyces fradiae NCIMB 8233: 2005. Characterisation of an aminotransferase involved in the formation of 2-deoxystreptamine. Fanglu Huang, Stephen F. Haydock, Tatiana Mironenko, Dieter Spiteller, Yanyan Li and Jonathan, B. Spencer. *Org. Biomol. Chem.*, 3:1410-1418.

Victor W. 1965. Metabolic Effects of the steroid antibiotic fusidic acid. *Brit Med. J.*, 1:1400-1404.

19

PhytopathoSys v1.0 – A Phytopathology Information System

M. Balakrishnan, R.C. Srivastava, Michael Imanual Jesse and M. Ramachandran

INTRODUCTION

Environment and ecological set up is maintained by the climatic conditions, presence and compositions of gases present in them, humidity level, temperature and the pollution. These conditions are responsible for the emerging of diseases in the environment. The diseases in plants, animals and human beings are due to the factors that affect the ecology. This also depends on the organism whether they are susceptible or resistant to the specific disease. This means the genetic level of the organism is an important factor.

Plant pathology or phytopathology is the study of diseases in plants. The disease evolution may be due to the climatic condition, humidity level, temperature and the genetic character of the plant and the pathogen. The parasitic adaptation of microorganisms to be plant pathogens includes acquisition of the following three abilities: 1) ability to enter into a plant, (2) ability to overcome the resistance of the host and (3) ability to evoke disease.

These abilities are essential to plant pathogens, and are not found in saprophytic microorganisms. Disease in a plant is an injurious physiological process caused by the continued irritation of a primary factor, exhibited through abnormal cellular activity and expressed in characteristic pathological conditions called symptoms and also the malfunctioning process caused by continuous irritation.

Plant pathogens have to enter into plants in order to derive food materials they require. They do this by one or more of the following four ways: entry through wounds, entry through natural openings, direct penetration and penetration through localized organs. After these entries the pathogen damages the internal system and leaves the plant diseased. Hence in order to create awareness among farmers and research people this information system is developed.

DAMAGE OF FIELD CROPS

The field crops include rice, sorghum, sesamum, Pulses and Sugarcane. These crops are susceptible to the diseases like Yellow and stem bores, Gundhi bug, Bacterial Leafblight, Sheath blight, Grain mould, charcoal rot, Root rot, Formation of Mosaic, Shoot bores, Red rot etc. and these diseases becoming a major challenge in the agricultural community.

DAMAGE OF VEGETABLE CROPS

The vegetable crops like tomato, Brinjal, Okra, Chilly, Cauliflower and Cabbage, Cucurbits, Legume and amaranthus are susceptible to the diseases fruit bores, bacterial wilt, Damping off, Fusarium wilt, Root not nematodes, Phomopsis blight, Yellow vein mosaic, Anthracnose and fruit rot, Leaf curl, Leaf spot, Downy mildew, Serpentine leaf miner, Rust, white rust etc. destroy the vegetable crops and reduce the economic value of those crops.

DAMAGE OF FRUIT CROPS

The cultivable fruits are banana, Mango, Citrus (Orange, Lime), Guava etc which are strictly infected by Rhizome weevil, Pseudo stem weevil, Banana Bunchy, Wilt, Fruit borer, Anthracnose, Leaf Miner, Black fly, fruit piercing moth, citrus butterfly, Canker, Spiralling White fly, Wilt etc. These destroy the crops at a maximum level and suppress the growth and production of fruit crops.

DAMAGE OF PLANTATION CROPS

The plantation crops include coconut and arecanut in Andaman and Nicobar group of Islands. They are mainly damaged by Rhinocerous beetle, Bud rot and Root grub

DAMAGE OF SPICES CROPS

The spices cultivation carried out in Andaman basically with Pepper, Clove and Ginger. These crops too subjected to damage by Pollu beetle, Top shoot borer, Foot rot, Stem borer, Shoot borer, Rhizome scale, soft rot etc.

ABOUT PHYTOPATHOSYS V1.0

The information system developed by Bioinformatics Centre, Central Agricultural Research Institute, Port Blair is a valuable system to reveal the details regarding the routes of infection, the microorganisms involved in the infection, from the host side the percentage and extent of damage. The treatment and control of disease both chemically and biologically. This system also gives information on the plants and the pathogens prevalent in Andaman and Nicobar Islands based on the research made.

METHODOLOGY

PhytopathoSysv1.0 is prepared using VB.net and MS SQL. VB.net is the advanced form VB6 and it gives the user a good working environment to know about the disease conditions and their preventive measures.

RESULTS

The System developed under VB.net consists of a Loading Screen at the beginning. When the user the Enter Button the loading bar starts running.After loading, the page takes you to the Home page which is built with a Menu bar with the menus File, Search by Pathogen, Search by host, Practices, Chemicals and About Us. The File menu consists of Open a file and Close the file or exit the entire program. Under Search By Pathogen tab, we given the user to select pathogen like Bacteria, Fungi, Virus and Pests. Under Search by Hosts tab, we have given user to select crops like field crops, vegetable crops, fruit crops, plantation crops, spices crops and tuber crops. Under Practices tab, one can select the type of practices to know about the methodology of practices. Under Chemicals tab, the type of chemicals used to control the pathogens and the disease. About us tab, takes the user to CARI Website and Contacts.

The user can also select the same from the check box in the Homepage. Simple Click the required *Check Box* and Clik '*Go*'.

For example, if we click the checkbox corresponding to Fungi and by clicking Go will take the user to the information page where one can find the combo box with the list of fungi. By selecting the query organism, the information on fungi including the name of the fungi, disease caused by the organisms on plants, the target crop, the extent of damage and the remark of the damage throughout the area. The microscopic picture of the organism is shown at the side of the window.

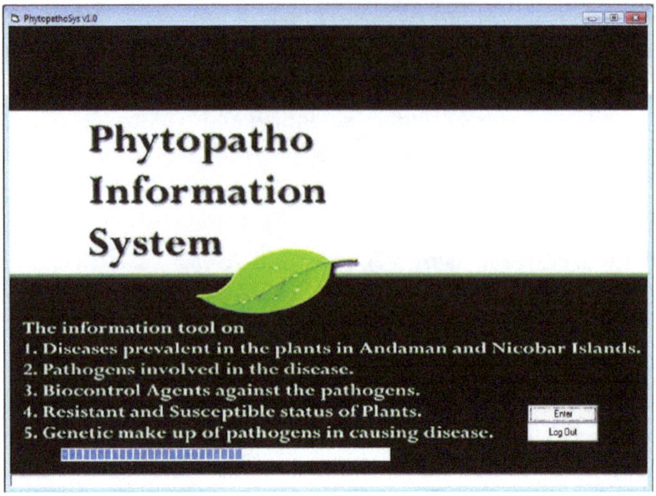
PhytopathoSys v1.0
Phytopatho
Information
System
The information tool on
1. Diseases prevalent in the plants in Andaman and Nicobar Islands.
2. Pathogens involved in the disease.
3. Biocontrol Agents against the pathogens.
4. Resistant and Susceptible status of Plants.
5. Genetic make up of pathogens in causing disease.
Enter
Log Out

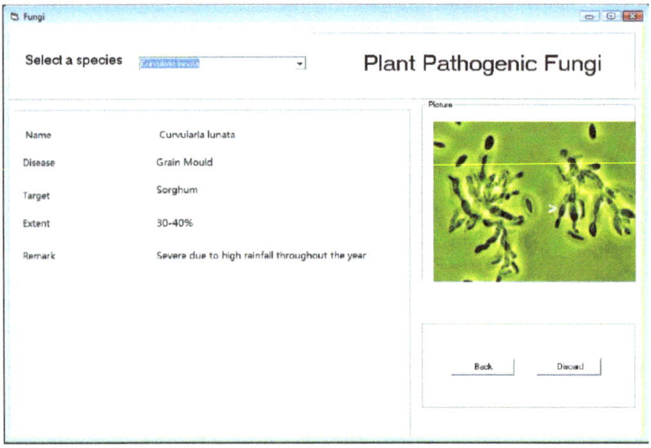
Fungi
Select a species
Plant Pathogenic Fungi
Picture
Name
Curvularia lunata
Disease
Grain Mould
Target
Sorghum
Extent
30-40%
Remark
Severe due to high rainfall throughout the year
Back

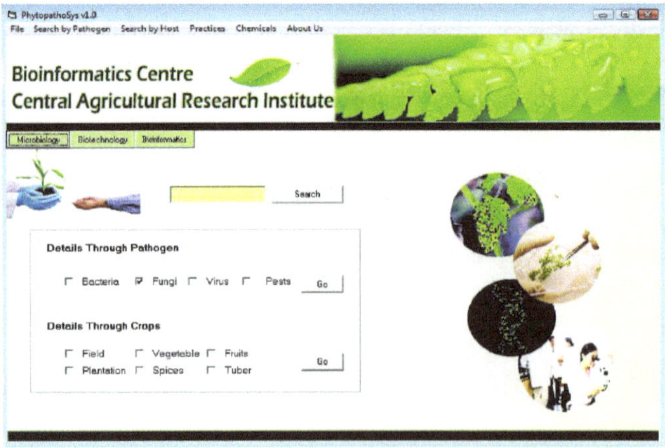
PhytopathoSys v1.0
File Search by Pathogen Search by Host Practices Chemicals About Us
Bioinformatics Centre
Central Agricultural Research Institute
Microbiology Biotechnology Bioinformatics
Search
Details Through Pathogen
Bacteria Fungi Virus Pests
Go
Details Through Crops
Field Vegetable Fruits
Plantation Spices Tuber
Go

REFERENCES

Jackson, R.W. 2009. Plant Pathogenic Bacteria: Genomics and Molecular Biology. Caister Academic Press. ISBN 978-1-904455-37-0.

Stewart, M. Gray, I. and Nanditta Banerjee. 2000. Mechanisms of Arthropod Transmission of Plant and Animal Viruses. *Microbiology and Molecular Biology Reviews*, 63:128–148.

Konstantin Kanyuka, Elaine Ward. and Michael, J. Adams. 1992. Polymyxa graminis and the cereal viruses it transmits: a research challenge. *Molecular Plant Pathology*, 4: 393-406.

Colson, P. Richet, H. Desnues, C. Balique and F. Moal, V. 2010. Pepper Mild Mottle Virus, a plant virus associated with specific immune responses, fever, abdominal pains, and Pruritus in humans, *PloS one* 5(4):e10041.

Choi, J.E. and Han, K.S. 1990. Bacterial soft rot and rhizome rot caused by Erwinia carotovora sub. sp. carotovora, Pseudomonas marginalis and P. solanacearum. Korean *J. Plant Path.* 6: 363-368.

George, N. Agrios. 1997. Plant Pathology fourth edition, Academic Press. New York.

Dianese, J.C. and Dristig, M.C.G. 1994. Strain characterization of Pseudomonas solanacearum based on membrane protein patterns. In: Bacterial wilt: Disease and its Causative agent, Pseudomonas solanacearum. (Eds. Hayward A.C. and Hartman, G.L.) CAB International. 113-171.

Dristig, M.C.G. and Dianese, J.C. 1990. Characterization of Pseudomonas solanacearum biovars based on membrane protein patterns. *Phytopathology* 80: 641-646.

Hara, H. and Ono, K. 1983. Ecological studies on the bacterial wilt of tobacco caused by Pseudomonassolanacearum E.F.Smith. *Bull. Okayama Tob. Exp. Stn.*, 42:127-138.

Mehan, V.K. 1995. Isolation and identification of P. solanacearum. In: Techniques for Diagnosis of Pseudomonas solanacearum• and for Resistance Screening against Groundnut Bacterial Wilt. (Eds. Mehan, V.K. and McDonald, D.) *ICRISAT Technical Manual*, 1:23-26.

Sarma, Y.R., Indrasenan, G. and Rohini, I. 1978. Bacterial wilt of ginger (Zingiber officinale). *Indian Arecanut Spices Cocoa J.* 2: 29-31.

Zehr, E.I. 1970. Cultural, physiological and biochemical properties of isolates to Philippine Pseudomonas solanacearum. *Philippine Phytopath.*, 6:29-43.

Kalen Delaney. 2009. Inside Microsoft SQL Server 2005. The Storage Engine. Microsoft Press. ISBN 0-7356-2105-5.

Christian Kleinerman. 2008. SQL Server 2008 for developers". Channel9. http://channel9.msdn.com/Showpost.aspx?postid=387069. Retrieved 2008-03-07.

Sherriff, Lucy. 2005. Real Software slams MS IsNot patent application. The Register. http://www.theregister.co.uk/2005/02/22/real_slams_ms_patent/. Retrieved 6 April 2009.

Taft, Darryl, K. 2005. Real Software Slams Microsofts Patent Effort. eWeek. http://www.eweek.com/article2/0,1759,1766949,00.asp. Retrieved 6 April 2009.

Vick, Paul, A. Barsan, Costica Corneliu, Silver and Amanda, K. 2003. United States Patent Application, Patent & Trademark Office, http://appft1.uspto.gov/netacgi/nph Parser. Retrieved 6 April 2009.

20

Multiple Gene Expression Profiles of Chemically Induced Parkinson Disease: Implication in Neuroprotection Mechanism Offered by Nicotine and Caffeine

Kavita Singh, M.P. Singh and M. Balakrishnan

ABSTRACT

Gene expression profiling holds tremendous promise for dissecting the regulatory mechanisms and transcriptional networks that underlie biological processes. Microarray expression analysis has become one of the most widely used functional genomics tools. Typical approaches to gene identification and functional characterization have and continue to involve protein characterization, peptide sequence determination, and identification of the corresponding DNA sequence. More recently, expressed sequence tags (ESTs), microarrays, large-scale gene expression (transcriptome) profiling, and associated informatics technologies are rapidly becoming commonplace in the sciences. These 'genomic' approaches typically take advantage of technologies for characterizing large numbers of nucleic acid sequences, bioinformatics, and the expanding collection of nucleic acid sequence data.

PESTICIDES INDUCED PARKINSON'S DISEASE

The cause of Parkinson's disease (PD) is unknown, but epidemiological studies suggest an association with pesticides and other environmental toxins with

Parkinson's disease(G.E. Meredith *et al.*, 2008; Terzioglu *et al.*, 2008). Herbicide rotenone causes highly selective nigrostriatal dopaminergic degeneration that is associated behaviorally with hypokinesia and rigidity. Experimental evidence supporting paraquatas a risk factor for Parkinson'sdisease is equivocal. Other agricultural chemicals, including dithiocarbamatefungicides such as manganese ethylenebisdithiocarbamate (maneb), are widely used in the same geographical regions as paraquatand also impact dopamine systems, suggesting that mixtures maybe more relevant etiological models. Previous reports suggest that pesticides exposure during pre-natal period can produce permanent and progressive lesions of the nigrostriatal dopamine system and can enhance the adult susceptibility to these pesticides. These results demonstrate the involvement of pesticide in induction of neurological disorders.

Epidemiological studies have shown an associationbetween risk of PD and environmental factors, such asexposure to pesticides. Combinedexposure to paraquat and maneb is consistentlyused as pesticide-induced animal model of PD in mouse (Singh *et al.*, 2006).Although the relevance of this model tohuman exposure and PD risk remains elusive, theneurotoxicity caused by paraquat and maneb is of greathealth concern, as farmers use these pesticides worldwide (Gorell *et al.*, 1998). As humans are exposed to a complex mixture of chemicals in their day-to-day life, establishing themechanism of neurodegeneration caused by pesticidesmay help in assessing the implications of environmental factors in PD (Brown *et al.*, 2006).

MICROARRAY IN NEUROLOGICAL DISORDERS

Parkinson's disease (PD) is the second most common neurodegenerative disorder after Alzheimer's disease. The disease is mainly characterized by selective degeneration of the nigrostriatal dopaminergic neurons resulting in tremor, bradykinesia, postural instability and rigidity (Thomas *et al.*, 2007; Galter *et al.*, 2008). Epidemiological and animal experimentations reveal the contribution of age, genetic and environmental factors in the etiology of PD. Among the environmental factors, pesticides and heavy metals play critical roles in the pathogenesis of PD (Tanner *et al.*, 1999; Ross *et al.*, 2007). Several environmental toxins have been used to develop rodent models to understand the biochemical and molecular mechanisms of PDpathogenesis and therapeutic interventions (Vieregge *et al.*, 1999; K. Wirdefeldt *et al.*, 2004). PD occurs sporadically; however, inheritance of single or multiple genes plays a critical role inits pathogenesis.

Despite detrimental effects of nicotine and caffeine, they offer neuroprotection by ameliorating neurodegenerative changes to some extent through various pathways. The inverse relationship between cigarette smoking/coffee drinking and risk of PD is reported in many studies (Patel *et al.*, 2009). Although several key genes have been identified in nicotine- and caffeine-mediated neuroprotection, the involvement of multiple genes leading to neuroprotection needs to be explored. Such approaches could be possible by using robotic tools, such as proteomics or microarray.

DNA microarray experiments can provide a powerful approach to reveal molecular mechanisms of neurodegeneration and neuroprotection by identifying simultaneous transcriptional regulation of thousands of genes. Differential gene expression profiling is often used to investigate the underlying molecular mechanism. The present study was therefore undertaken to delineate the involvement of multiple genes and their possible roles in nicotine- and caffeine-mediated neuroprotection.

METHODOLOGY

Animal treatment and collection of brain tissue

Male Swiss albino mice were treated with MPTP to induce Parkinson's disease and control animals were treated intraperitoneally with normal saline. The nicotine or caffeine doses are given to offer neuroprotection.

RNA ISOLATION

Total RNA was isolated from the striatum using Tri reagent (Sigma–Aldrich, St. Louis, USA) following manufacturer's instructions. RNA quality was assessed by measuring the ratio of optical density at 260/280 and was quantified by taking absorbance at 260 nm. Band density ratio of 28S/18S checked the integrity of RNA after formaldehyde agarose gel electrophoresis. The ratio should be never less than 2:1.

cDNA LABELING AND MICROARRAY EXPERIMENTATION

Pre-arrayed mouse 15k slides used in the present study were procured from University Health Network, Canada. In brief, these slides were developed from the original clone sets developed by National Institute of Aging (NIA), PCR amplified, purified and spotted onto CMT-GAP II slides (complete information available at http://www.microarrays.ca). Comparative transcription profiling was performed using a two-color dye hybridization protocol. In brief, RNA was reverse transcribed into labeled cDNA by incorporating cyanine 3- (cy3) or cyanine 5-labeled (cy5) dUTP. Labeled cDNA was separated from unincorporated nucleotides and further concentrated by evaporation under vacuum.

CDNA HYBRIDIZATION

The labeled cDNA was quantified and a mixture of equal amounts of cyanine-3 and cyanine 5-labeled cDNA was heated at 95^{0}C for 5 min and hybridized with expressed sequence tags spotted onto mouse 15k array. The hybridization was performed at 42°C for 18 h and hybridized slides were stringency washed at room temperature. Three experiments were performed for each group and similarly

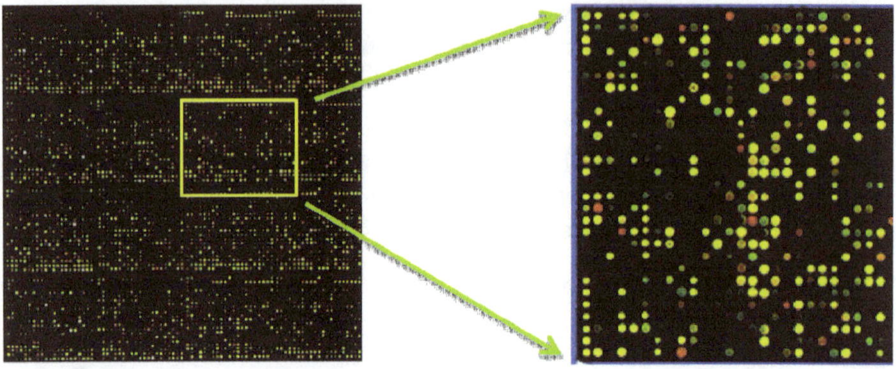

(a) Microarray Scanned Slide (b) Magnified Grid

Fig 1: The scanned image of Mouse 15K array slide in (a) and magnified grid of the complete arrayed chip (b),red indicates up-regulated, green represents down-regulated transcripts and yellow indicates that both the transcripts are equally expressed.

three dye swap technical replicates were also performed to minimize variability arising due to the incorporation of dye and to avoid other technical problems. Hybridized arrays were scanned using a laser scanner (GeneTac LS IV, Genomic Solutions, USA) under green (550 nm for cyanine-3) and red laser (675 nm for cyanine-5). The analyses of scanned output files were performed by Array Vision 8.0 (GE Healthcare, Europe) software.

MICROARRAY DATA ANALYSIS

A total of three normal and three dye swap replicate experiments were performed for three independent sets of experiments. Using one dye for labeling of cDNA in control and another for treated groups three sets of normal experiments were performed. For another three sets of dye swap replicates, the dye used to label control was used to label treated cDNA and vice versa. Three dye swap technical replicates of the same RNA samples that were used for three independent normal experiments were performed to minimize the inconsistency and technical variability arising due to incorporation of dyes. After scanning, the array images were imported into a software program Array Vision 8.0 (GE Health-care, Europe) for spot detection and quantification of hybridization signals. Each spot was defined by positioning of a grid of circles over the array image and spot density for each gene was measured using quality control measures defined in the manual. Although the grids were pre-defined, they were adjusted manually as well to ensure the optimal spot recognition. Spots were individually quantified and local backgrounds were calculated from the corners between the spots. Signal intensity for each spot was determined by subtracting the background from intensity. Spots with high background noise and not circular in shape were ignored in data analyses.

DATA NORMALIZATION

Expression data were normalized using locally weighted scatter plot smoothing (LOWESS) method. Data were filtered to avoid false positive spots by excluding genes with a signal-to-noise ratio <3. Data files generated by Array Vision 8.0 software were entered into a custom database. As each slide comprised of duplicate transcripts, averaging the intensities of the duplicates did in-slide replicate analysis. The fold changes for the differentially expressed genes were calculated from the ratios of intensities of the two test groups. These results were verified using image analysis software (GeneTac Integrator Version 4.0, Genomic Solutions, USA). Genes showing consistency in fold changes between normal experiments and dye swap replicates were selected.

LOWESS NORMALIZATION

In experiments where two fluorescent dyes (red and green) have been used, intensity-dependent variation in dye bias may introduce spurious variations in the collected data. Lowess normalization merges two-color data, applying a smoothing adjustment that removes such variation.

LOWESS NORMALIZATION CHARACTERISTICS

- Lowess normalization may be applied to a ***two-color*** array expression dataset.
- All samples in the dataset are corrected independently.
- Lowess normalization can be applied to complete or incomplete datasets. If either the red or green intensity value is missing for a certain gene, there will be a missing value at the corresponding position in the log-ratio table which is generated.

LOWESS NORMALIZATION METHOD

Lowess normalization assumes that the dye bias appears to be dependent on spot intensity. The adjusted ratio is computed by:

log(R/G) -> log(R/G) – c(A)

where c(A) is the Lowess fit to the log(R/G) vs log(sqrt(R*G)) plot.

If green has been chosen as the treatment dye and red as the control dye, then R and G are reversed in the above formula. Treatment and control dyes are designated when the data is imported into GeneLinker™.

Lowess regression, or locally weighted least squares regression, is a technique for fitting a smoothing curve to a dataset. The degree of smoothing is determined by the window width parameter. A larger window width results in a smoother curve, a smaller window results in more local variation.

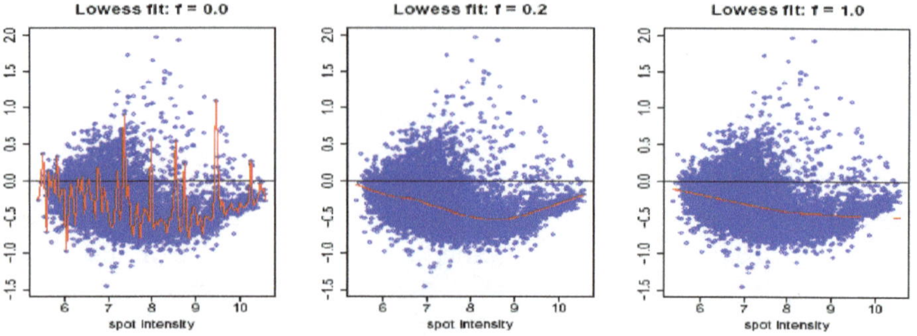

Fig 2: Successful completion of the normalization, a new dataset with the Lowess-corrected (R/G) values (or (G/R) if appropriate) is stored in the repository.

HIERARCHICAL CLUSTERING

As mentioned above, raw expression data were filtered using LOWESS-normalization and background correction. Hierarchical clustering was through average linkage for identification of genes with similar expression patterns using Cluster version 2.11software and tree view was generated using TreeView version 1.60 software. A possible involvement of molecular pathways was speculated using the datasets obtained for some of the differentially expressed genes and a representative gene expression dataset using online freely available version of the software GenMAPP version 2.1.

Inferences for functional annotation of genes were obtained from following databases:

NIA (http://Igsun.grc.nia.nlm.gov)

NCBI (http://ncbi.nlm.nih.gov/)

SOURCE (http://smd.stanford.edu/cgibin/source/sourceBatchSearch)

SwissProt (http://www.expasy.org/sprot/)

QUANTITATIVE REAL-TIME POLYMERASE CHAIN REACTION

To validate the expression patterns of genes obtainedusing microarray, eight representative genes were subjected to quantitative RT-PCR (qRT-PCR). RNA (5 mg) wasreverse-transcribed into cDNA using oligo-dT primers. Primersequences were identified from the online PrimerBank database(http://pga.mgh.harvard.edu/primerbank/index.html.) and synthesized accordingly . Desired segments oftoll-interleukin-1 receptor domain-containing adaptor protein,nuclear protein-1, cathepsin B, interleukin-4 receptor, caspase 9,complement component 1, heat shock protein-1 and cytochromec-oxidase-VIIc were amplified using 100 ng cDNA, 400 nM of eachprimer and SYBR green 2X-master mix in reaction mixture (25 ml).The mouse glyceraldehyde-3-phosphate dehydrogenase (GAPDH)gene was amplified as an endogenous control with the representative genes. Levels of qRT-PCR products were measured using SYBR green fluorescence in RT-PCR system (Applied Biosystems 7900, USA). A cycle threshold (Ct) was assigned in the beginningof logarithmic phase of amplification and difference in Ct values ofthe control and treated groups were used to determine the relativeexpression. Each reaction was subjected to melting point analysisto confirm single amplified products.

STATISTICAL ANALYSIS

The data are expressed as means ± standard error of means(S.E.M.) (n = 6; in 3 sets of experiments, controls were labeled withCye 3 and treated with Cye 5 dyes while in the remaining 3 sets,dyes were swapped). One-way analysis of variance (ANOVA), followed by Newman–Keuls post-test and student's t-test were usedfor comparisons. The differences were considered statistically significant, when 'P' value was <0.05. The results of microarray baseddifferential gene expression pattern are expressed as fold changesof signal intensities of treated versus (vs.) control. qRT-PCR data areexpressed in terms of fold change with respect to the endogenouscontrol.

RESULTS AND DISCUSSION

Representation of Gene Expression Data

Microarray analysis of MPTP-intoxicated mouse striatum with or without nicotine or caffeine treatment showed significant differential expression of several transcripts associated with ubiquitin–proteasome system, mitochondrial respiratory chain, energy metabolism, inflammation, transcription, apoptosis, oxidative stress, cell cycle regulation and neurodegeneration. MPTP treatment alters the expression of 772 total transcripts more than 2 folds as compared with control that comprised of 345 up-regulated and 427 down-regulated transcripts. The expression pattern of nicotine pre-treated MPTP exposed animals showed the differential expression of 540 transcripts (130 up-regulated, 410 down-regulated) as compared with control. 387 transcripts (207 up-regulated, 180 down-regulated) were also differentially expressed in caffeine pre-treated MPTP exposed animals as compared with control. Only those differentially expressed genes, which were significantly altered following MPTP exposure and whose expression was affected with nicotine and caffeine, were chosen to assess their possible contribution in neurodegeneration/ protection. However, MPTP, nicotine + MPTP and caffeine + MPTP altered several genes independently. The comparative gene expression patterns exhibited differential expression of 772 transcripts after MPTP exposure, among which 113 transcripts were altered by nicotine and caffeine.

Table 1:

Biological pathways	Clone ID	Gene name	Symbol	Fold Change	
				MPTP	Caffeine + MPTP
Apoptosis	H3090D01	mitogen activated protein kinase kinase kinase 12	Map3k12	5.55 ± 0.87	-6.56 ± 1.05
	H3001F01	ephrin A1	Efnal	-3.00 ± 0.23	7.90 ± 0.51
Cell division	H3010F01	cyclin F	Ccnf	15.37 ± 2.51	7.28 ± 1.27
	H3144D03	helicase, lymphoid specific	Hell	5.88 ± 0.38	2.36 ± 1.29
Cytoskeleton	H3115F07	phosphatidylinositol glycan, classC	Pigc	14.20 ± 0.72	4.63 ± 1.08
Energy metabolic	H3018G08	galactose-4-epimerase, process UDP	Gale	5.53 ± 1.21	16.15 ± 1.51
	H3098H02	discarbonyl L-xylulose reductase	Dexr	-7.95 ± 1.04	3.71 ± 1.02
	H3040H10	NADH delrydrogenase (ubiquinone) Fe-S protein 4	Ndufs4	-7.03 ± 0.86	4.54 ± 1.16
	H3025FF11	glyceraldehyde-3-phosphate dehydrogenase	Gapd	-4.11 ± 0.58	-2.36 ± 0.16
	H3137F11	uroporphyrinogen Ill synthase	Uros	-2.11 ± 0.15	2.02 ± 0.10
Lon transport	H3019H11	ATPase, H + transporting V0 subunit C	AtpGv0c	-5.20 ± 0.39	2.99 ± 0.199
	H3019H11	ATPase, H+ trans porting, V1 subunit B, iso form 2	Atp6vl b2	-3.55 ± 0.70	3.50 ± 0.21
Neuron myelination	H3129H08	Fyn proto-oncogene	Fyn	-6.02 ± 0.60	-3.76 ± 0.14
Oxid oreductase activity	H3099G11	similar to chromosome1 open reading frame 27	LOCC226499	02.42 ± 0.15	4.14 ± 0.47

(Contd.)

Biological pathways	Clone ID	Gene name	Symbol	Fold Change	
				MPTP	Caffeine + MPT
Protein modification	H3157A07	chaperone, ABC1 activity of lx1 complex like (S. pombe)	Cabc1	-5.55 ± 1.16	11.87 ± 1.63
	H3059F01	ubiquitin carboxy-terminal hydrolase11	Uchl1	-6.48 ± 0.71	-3.74 ± 0.98
Signal transduction	H3118H10	melanocortin 2 receptor	Mc2r	-4.68 ± 0.22	9.70 ± 1.25
Transcription	H3005F02	heterogeneous nuclear	Hnrpc	3.10 ± 0.31	5.62 ± 1.09
Ribonucleoprotein C	H3112B01	Sin3 associated polypeptide 18	Sap18	11.76 ± 2.03	7.96 ± 1.32
Translation	H3126H10	eukaryotic translation elongation factor 1 alpha 1	Eef1a1	9.41 ± 1.73	3.29 ± 0.29
	H3129B11	eukaryotic translation elongation factor 2	Eef2	9.80 ± 1.92	2.78 ± 0.08

Table 1. A table showing some of the differentially expressed transcripts related to various biological pathways along with their clone identity numbers, gene abbreviations and relative fold changes in expression after MPTP and caffeine +MPTP treatment as compared with saline treated controls. Fold changes of down-regulated genes are shown as negative and up-regulated genes as positive values. The data are expressed as means±S.E.M. Only those transcripts are shown in table whose expressions were modulated after caffeine exposure but not after nicotine exposure.

Table 2:

Biological Pathways	Clone ID	Gene name	Symbol	Fold Change	
				MPTP	Nicotine +MPTP
Apoptosis	H3124F07	RAB22, member RAS oncogene family	Rab2	24.12 ± 2.64	3.31 ± 0.31
	H3021H09	CD40ligand activated	Clast3-pending	14.02 ± 1.85	-2.32 ± 0.21
	H3103E12	Bc12-like 10	Bcl2110	-3.33 ± 0.31	-2.24 ± 0.08
	H3105D09	apoptosis-associated speck-like protein containing a CARD	Asc-pending	4.23 ± 0.94	2.31 ± 0.17
	H3131G02	caspase 12	Casp12	20.89 ± 3.08	5.02 ± 0.20
	H3138C12	activity-dependent neuro protective	Adnp	-3.52 ± 0.09	3.02 ± 0.24
	H3017F06	mitogen activated protein kinase 1	Mapk1	16.42 ± 0.85	-3.50 ± 0.10
	H3005G09	growth factor receptor bound 2-associated protein 1	Gab1	6.69 ± 0.52	-2.51 ± 0.06
	H3147D12	apoptosis inhibitor 5	Api5	-5.57 ± 0.53	-2.05 ± 0.24
Cell cycle	H3018F01	protein (peptidyl-prolyl cis/trans isomerase) NIMA-interacting 1	Pin1	18.93 ± 0.68	-2.5 ± 0.04
Cell division	H3026D04	cyclin D2	Ccnd2	2.16 ± 0.50	-2.2 ± 0.32
	H3027G03	cyclin B1	Ccnb1	2.28 ± 0.22	-2.26 ± 0.12
	H3158C12	calmodulin binding protein1	Calmbp1	-7.34 ± 0.98	-5.24 ± 0.12
	H3147F11	cell division cycle 42 homolog (S.cerevisiae)	Cdc42	10.40 ± 1.49	-2.7 ± 0.01
	H3139C03	CD9 antigen	Cd9	20.81 ± 2.00	-3.98 ± 0.11
Energy metabolism	H3140D12	electron transferring flavor protein alpha polypeptide	Etfa	-5.80 ± 0.70	-3.96 ± 0.08
	H3025H1	glyceral dehyde-3-phosphate dehydrogenase	Gapd	-4.10 ± 0.22	-2.07 ± 0.04
Inflammatory response	H3012H07	CD44 antigen	Cd44	20.73 ± 3.22	5.46 ± 0.36
	H3004D02	phospholipase A2, activating protein	Plaa	11.78 ± 1.60	3.39 ± 0.51

(Contd.)

Biological Pathways	Clone ID	Gene name	Symbol	Fold Change	
				MPTP	Nicotine +MPTP
	H3031E01	tissue inhibitor of metalloproteinase 3	Timp3	-2.34 ± 0.12	3.70 ± 0.03
Metabolic	H3112B07	acetyl-Coenzyme A process acyltransferase 2 (mitochondrial 3-oxoacyl-Coenzyme A thiolase)	Acaa2	-2.10 ± 0.06	-2.72 ± 0.07
	H3147F06	malic enzyme, supernatant	Mod1	-5.39 ± 0.03	-3.79 ± 0.09
	H3152H05	ATPase, class II, type 9A	Atp9a	-3.62 ± 0.24	-2.91 ±0.17
Neuron specific	H3150D06	Parkinson disease (autosomal recessive,early onsef) 7	Park7	16.08 ± 1.19	4.43 ± 0.34
	H3042F07	synuclein, gamma	Sncg	22.90 ± 2.72	-2.27 ± 0.19
	H3152B03	glutanrate receptor, ionotropic, delta 2	Grid2	8.82 ± 0.74	-1.52 ± 0.01
Oxidation reduction	H3140D07	peroxiredoxin 3	Prdx3	10.02 ± 1.30	-5.62 ± 0.44
	H3094D01	cytochrome P450, family11, subfamilya polypeptide1	Cyp11a1	-6.70 ± 0.24	2.31 ± 0.36
	H3153F08	phosphatidylserine receptor	Ptdsr	13.09 ± 2.07	-4.04 ± 0.24
	H3119G08	glutathione S-transferase, alpha 4	Gsta4	8.26 ± 0.56	3.71 ± 0.52
Protein modification	H3129G09	ubiquitin carboxy-terminal hydrolase L1	Uchl1	-7.90 ± 1.08	-3.17 ± 0.01
	H3028F01	Brcal associated protein 1	Bap1	11.30 ± 0.87	-3.28 ± 0.19
	H3022G04	proteasome (prosome, macropain) 26S subunit, ATPase2	Psmc2	-2.97 ± 0.5	-2.81 ±0.03
	H3028F04	cathepsin L	Ctsl	12.81 ± 1.22	-3.28 ± 0.07
Response to DNA damage	H3051F03	calcium and integrin binding 1 (calmyrin)	Cib 1	5.85 ± 0.82	3.21 ± 0.05
	H3097F05	protein kinase, AMP activated, beta 1 non-catalytic subunit	Prkab 1	-5.73 ± 0.81	-2.41 ± 0.36
	H3005A09	glutamate-cysteine ligase, catalytic subunit	Gclc	-2.98 ± 0.07	-2.10 ± 0.04

(Contd.)

Biological Pathways	Clone ID	Gene name	Symbol	Fold Change	
				MPTP	Nicotine +MPTP
Signal transduction	H3076D03	G protein-coupled receptor,family C.group 5, member C	Gprc5c	9.33 ± 0.62	5.19 ± 0.27
	H3152B05	homeodomain interacting protein kinase 1	Hipk1	4.52 ± 0.51	-2.91 ± 0.09
	H3106F07	guanine nucleotide binding protein, alpha inhibiting 2	Gnai2	-3.93 ± 0.22	-2.53 ± 0.45
Structural	H3133E09	synaptogyrin2	syngr2	-3.76 ± 0.54	-3.15 + 0.02
	H3140E06	CED151 antigen	Cd151	-3.02 ± 0.21	-2.32 ± 0.06
	H3151F07	procollagen, type VI, alpha 1	Col6a1	6.50 ± 0.92	2.32 ± 0.07
	H3040B09	mitocliondrial ribosomal protein L10	Mrpl10	-4.92 ± 0.48	-2.78 ± 0.28
Transcription	H3005F02	heterogeneous nuclear ribonucleoprotein C	Hnrpc	-13.12 ± 0.21	-4.14 ± 0.08
	H3046B11	GATA binding protein 6	Gata6	11.66 ± 1.06	2.51 ± 0.04s
	H3050D12	histone acetyltransferase	Morf-pending	-3.98 ± 0.40	-3.18 ± 0.76
	H3092E09	early growth response 1	Egr1	-5.20 ± 0.23	-2.29 ± 0.14
Translation	H3029H03	eukaryotic translation initiation factor 2B	Eil2b	-5.80 ± 0.15	-2.95 ± 0.04
Transport	H3077F08	ATPase, H+transporting, lysosomal V0 subunit A isoform 4	Atp6v0a4	-10.14 ± 0.47	-3.41 ± 0.01
	H3019C04	vacuolar protein sorting 28 (yeast)	Vps28	-11.23 ± 1.22	-2.17 ± 0.04
	H3152C07	ADP-ribosylation factor 3	Arf3	-7.06 ± 0.28	-4.69 ± 2.28
	H3136A06	chloride channel calcium activated 2	Clca2	-3.12 ± 0.12	4.80 ± 0.79
	H3136H01	neuroglobin	Ngb	-3.43 ± 0.08	-2.09 ± 0.36

Table 2. A table showing some of the differentially expressed transcripts related to various biological pathways along with their clone identity numbers, gene abbreviations and relative fold changes in their expression after MPTP and nicotine +MPTP treatment as compared with saline treated controls. Fold changes are expressed as means±S.E.M. The down-regulated genes are shown as negative and up-regulated genes as positive values. Only those transcripts are shown in table whose expressions were modulated after nicotine exposure but not after caffeine exposure.

GENE CLUSTERING AND FUNCTIONAL CHARACTERIZATION

Comparative evaluation of expression patterns of MPTP, nicotine + MPTP and caffeine + MPTP co-treated animals exhibited a significant alteration in the expression of transcripts associated with many pathways incorporating ubiquitin–proteasome system, mitochondrial respiratory chain, energy metabolism, inflammation, transcription, apoptosis, oxidative stress, cell cycle regulation, dopamine signaling, ribosomal proteins and neurodegeneration. Only those transcripts were selected for clustering, whose differential expression level was

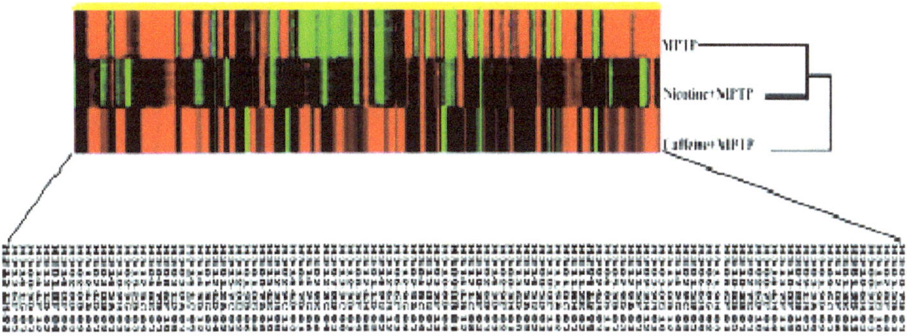

Fig 3: Clustered display of data obtained from microarray experimentation of the striatum of control, MPTP, caffeine +MPTP and nicotine +MPTP-treated mice. All measurements are relative to control. Only those transcripts were selected for the analysis, whose average expression level was altered (>2 fold) by MPTP treatment and was modulated by caffeine and nicotine pre-treatment. Red indicates up-regulated and green represents down-regulated transcripts. The transcript/ clone/gene ID are also mentioned in the figure.

e" 2.0 folds in MPTP-treated group as compared with control and were modulated by nicotine and caffeine.

QUANTITATIVE REAL TIME-PCR

Quantitative RT-PCRthe expression levels of toll-interleukin-1 receptor domain-containing adaptor protein, nuclear protein-1, cathepsin B,interleukin-4 receptor, caspase 9, complement component-1, heatshock protein-1 and cytochrome c-oxidase-VIIc validated thetrends of gene expressions as obtained in microarray experimentations (Fig. 4).A possible involvement of a few molecular events is speculatedusing the datasets obtained for some of the differentially expressedgenes

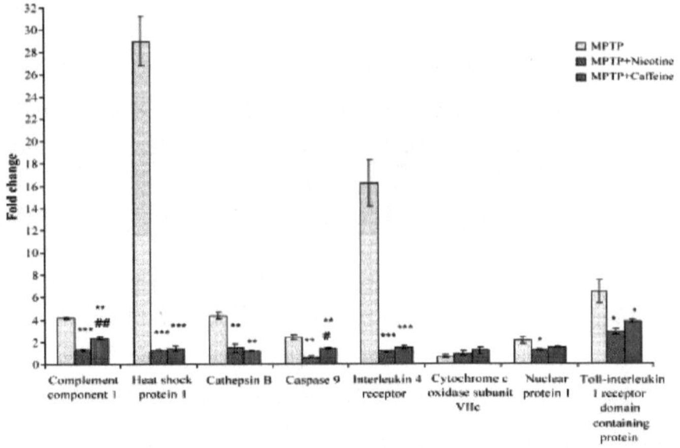

Fig. 4: qRT-PCR validation of some differentially expressed transcripts. Bar diagrams showing differential expressions of complement component-1, heat shock protein-1, cathepsin B, caspase 9, interleukin-4 receptor, cytochrome c-oxidase-VIIc subunit, nuclear protein-1and toll-interleukin-1 receptor domain-containing adaptor proteintranscripts in MPTP-treated mouse brain in presence or absence of nicotine or caffeine as compared with controls.

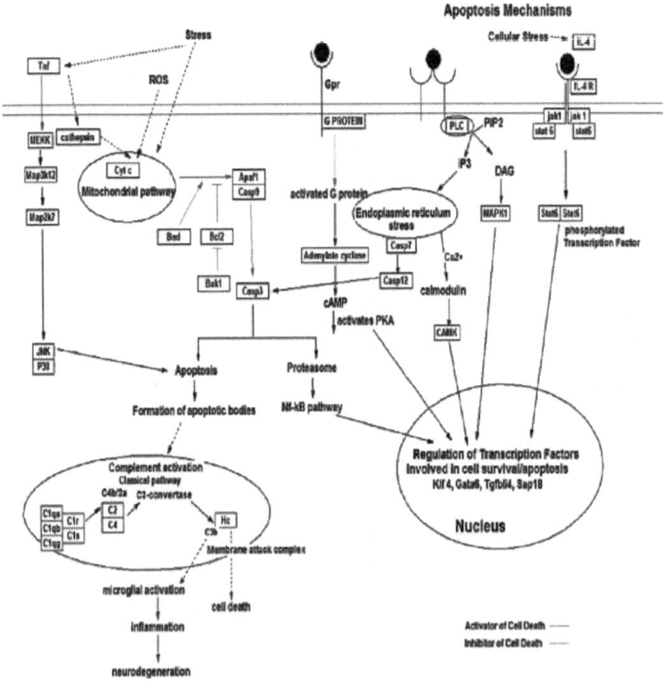

Fig. 5: Speculated signaling pathway involved in MPTP-mediated neurodegeneration and nicotine/caffeine-mediated neuroprotection (based on online freely availableversion of the software GenMAPP version 2.1).

using online freely available version of the software GenMAPP version 2.1 (Fig. 5).

CONCLUSION

The results obtained from differential gene expression profiles suggest that various molecular pathways, including ubiquitin proteasome system, mitochondrial respiratory chain, energy metabolism, inflammation, apoptosis, oxidative stress, cell cycle regulation are possibly involved in MPTP-mediated neurotoxicity and nicotine- and caffeine-mediated neuroprotection. However, involvement of these pathways needs to be confirmed one by one using separate molecular tools to enable a more definitive conclusion. Additionally, the study reveals that nicotine and caffeine share some common pathways of neuroprotection despite many differences.

REFERENCES

Brown, T.P., Rumsby, P. C., Capleton, A. C., Rushton, L., and Levy, L. S. 2006. Pesticides and Parkinson's disease-is there alink? *Environmental Health Perspectives*, 114:156-164.

Dahlquist, K.D., Salomonis, N., Vranizan, K., Lawlor, S.C., Conklin, B.R. 2002. GenMAPP, a new tool for viewing and analyzing microarray data on biological pathways, *Nat. Genet.* 31:19–20.

Galter, D. 2008. Parkinson's Disease: Recent Progress, *FEBS J.* 275:1369.

Gorell, J.M., Johnson, C.C., Rybicki, B.A., Peterson, E. L. and Richardson, R.J. 1998. The risk of Parkinson's disease withexposure to pesticides, farming, well water, and rural living. *Neurology*, 50:1346-1350.

Meredith, G.E., Sonsalla, P.K., Chesselet, M.F. 2008. Animal models of Parkinson's disease progression, *Acta Neuropathol.* 115:385-398.

Patel, S., Singh, K., Singh, S., Singh, M.P. 2008. Gene expression profiles of mouse striatum in control and maneb + paraquat-induced Parkinson's disease phenotype: validation of differentially expressed energy metabolizing transcripts, *Mol. Biotechnol.* 40:59–68.

Ross, C.A., Smith, W.W. 2007. Gene-environment interactions in Parkinson's disease, Parkinsonism *Relat. Disord.* 13 (Suppl. 3):S309-315.

Singh, M.P., Patel, S., Dikshit, M., and Gupta, Y.K. 2006. Contribution of genomics and proteomics in understanding therole of modifying factors in Parkinson's disease. Indian *Journal of Biochemistry Biophysics*, 43:69-81.

Singh, S., Singh, K., Gupta, S.P., Patel, D.K., Singh, V.K., Singh, R.K., Singh, M.P. 2009. Effect of caffeine on the expression of cytochrome P450 1A2, adenosine A2Areceptor and dopamine transporter in control and 1-methyl 4-phenyl 1,2,3,6-tetrahydropyridine treated mouse striatum. *Brain Res.* 1283:115-126.

Tanner, C.M., Ottman, R., Goldman, S.M., Ellenberg, J., Chan, P., Mayeux, R., Langston, J.W. 1999. Parkinson disease in twins: an etiologic study, *JAMA*, 281:341-346.

Terzioglu, M., Galter, D., 2008. Parkinson's disease: genetic versus toxin-induced rodent models, *FEBS J.* 275:1384-1391.

Thomas, B., Beal, M.F. 2007. Parkinson's disease, Hum. Mol. Genet. 16 (Spec No. 2): R183-194.

Vieregge, P., Hagenah, J., Heberlein, I., Klein, C., Ludin, H.P. 1999. Parkinson's disease in twins: a follow-up study, *Neurology*, 53:566–572.

Wirdefeldt, K., Gatz, M., Schalling, M., Pedersen, N.L. 2004. No evidence for heritability of Parkinson disease in Swedish twins, *Neurology*, 63:305-311.

21

Development of Databases on Freshwater Aquaculture

A.K. Roy and Nibedita Jena

Keeping pace with latest technologies Central Institute of Freshwater Aquaculture (CIFA) has attempted to develop the following three databases on the following aspects for quick retrieval of information resources. The brief description of the databases is presented below for easy comprehension.

CD-ROM DATABASE ON SOCIO-ECONOMIC ASPECTS OF CARP CULTURE IN KOLLERU LAKE (PICTORIAL)

In the era of 21st century world is progressing fast towards the electronic media of communication and the real revolution has been brought about in the information technology. It has revolutionized the collection, storage and retrieval of information very speedily and accurately. One of the IT tools like CD-ROM (Compact Disc Read Only Memory) plays an important role in storing the huge data and can be easily carried to remote places. Since it is lightweight, pocket sized and can hold about 660 megabytes of data which is equivalent to 3 lakh pages. CD-ROM is a high capacity optical disc. It is 12 cm diameter, made of durable plastic with a reflective metal coating and lacquered surface. Considering all these advantages an attempt has been made to develop a database on socio-economic aspects of carp culture in Kolleru Lake which is immensely helpful to all scientific and farming community (Roy *et al.*, 2004).

BACKGROUND OF KOLLERU LAKE

Kolleru Lake is an extensive shallow depression lying in between West Godavari and Krishna districts of Andhra Pradesh. It is the largest freshwater lake in India covering an area of about 2,96,640 ha. The population of the Kolleru Lake mainly comprised of fishermen community distributed in 50 bed (Island) habitation and 98 belt (border) habitations spread over 70 revenue villages. The major occupation in bed habitations is fishing. Prior to 1970's it was only the area of capture fisheries and paddy cultivation, which was followed by culture fisheries in 1980's. At present aquaculture has taken the shape of an industry and it also plays a vital role in rural development.

MATERIALS AND METHODS

For assessment of socio economic conditions of fish farmers a wide range of data are needed. For this purpose several field visits during 1998-2001 were undertaken into different villages in different mandals (blocks) in Kolleru region. Detailed information regarding aquaculture practices, socio-economic conditions, ancillary industries were collected by personal interview method. Photographs are also taken covering all the above said activities. To handle this huge data an attempt has been made to develop a database.

DATABASE DESIGNING

As it is well known that a well-designed database provides quick and accurate access to information. For this purpose all the pictorials, about 200 are arranged under the following heads

S. No	Database Heads
1.	View of Kolleru lake
2.	Traditional capture fishery
3.	Construction and conversion
4.	Weed control
5.	Pond views
6.	Water management
7.	Fertilization
8.	Seed rearing
9.	Feeding
10.	Disease control
11.	Watch and ward
12.	Integrated fish farming systems
13.	Harvesting
14.	Ancillary industries
15.	Socio-economic conditions
16.	Prawn farming
17.	Marketing

Initially all the photos were scanned using HP Scanjet 5100c. Then these photos are linked using Present Pro software. To execute/activate the database and the above head need to be selected from the homepage and it will link to the respective items. On each page the buttons used for navigation are next, previous, quit and index.

Previous button: links to previous page.

Next button: links to next page.

Index button: links to the homepage.

Exit button: quit from database.

Brief descriptions of the heads are mentioned below.

VIEW OF KOLLERU LAKE

It covers origin of the lake, different types of available weeds, canals, bird sanctuary and passage to island villages.

CAPTURE FISHERIES

The thick mass of floating vegetation restricts the movement of conventional boats. Hence only the round bottomed canes made out from palm tree are used. Usually morning time fishermen go inside the lake by using these boats for fishing. They carry traps for capturing air breathing fishes. These traps are made of split bamboo. These traps are placed at different places and collected in evening time.

POND CONSTRUCTION AND RENOVATION

Now-a-days farmers use bulldozers and proclainers for earthwork, to construct and renovate the tank which otherwise might involve not only a large number of labourers but also long time. Use of machines has made this work very easy, quick and convenient.

WEED MANAGEMENT

Control of excessive weed growth is an important measure for the resource management. At present weeds are removed manually and by chemical application. For killing unwanted predatory weeds endosulphan is used at the rate of 2.5 l/ha. Information regarding weed management covered in this head.

POND VIEWS

Different sizes of pond viz., nursery ponds, rearing ponds, stocking ponds are recorded. Fish farmers in Kolleru Lake prefer larger size water bodies, as these are conducive for better growth of fish leading to higher productivity. The farm size and aquaculture productivity found positively correlated. The size of the stocking pond ranges between 1-101 ha and water depth ranges between 6-10 feet. Most of the ponds are well connected by infrastructure facilities like roads, canals and inlet-outlet facilities. Pond dykes are used for growing coconut trees and vegetables, which adds to the extra income. Most of the ponds are found to be in rectangular shape.

WATER MANAGEMENT

Water is the main source of aquaculture. It plays a very important role in deciding the size and depth of the pond. Since water is available abundantly, ponds in Kolleru are very big in size. As the ponds are filled from canal water the entry of extraneous fish and other organisms in the egg or larval stages with water are avoided by tying small meshed screens to inlet and outlet pipes. Most of the villages are electrified, if there is no electricity facility diesel pump sets are seen in use. Pump houses were also constructed near the pond site. The loss of water from a fish tank through seepage, evaporation etc., is being compensated through regular pumping of water into the tank.

FERTILIZATION

Before applying fertilizer and lime, the ponds are initially treated with endosulphan to kill unwanted fish. After 10-15 days of chemical application, ponds are fertilized. Earlier organic fertilizer was applied to the pond at the rate of 4 t/ha at the time of preparation of pond and also once in every month depending upon the plankton growth. Most of the farmers are using poultry litter, as it is cheaply available due to existence of large number of poultry farms. Farmers are also using urea, DAP and SSP to increase the primary production of the pond.

SEED REARING AND STOCKING

At present species combination confined to rohu and catla only, due to change in consumer preference and market demand. Existence of hatcheries in large scale eased the availability of seed to the farmers round the year. Nursery ponds are maintained at 1-2 ft. depth. Initially ponds are desilted, dried and then lime applied which gives conducive atmosphere for the fry to grow. Fry of 4" size stocked at the rate of 30,000-70,000/acre. Fry is available abundantly during the months of July to January thereafter for the rest of the period the seed production is comparatively less. These seeds are transported in lorries/tractors using syntax tanks filled with water and aeration provided by using oxygen cylinders. These fish is stocked in stocking ponds at the rate of 7000-10000/ha out of which rohu alone constitute 80% of the total stock.

SUPPLEMENTARY FEED

Feed plays a vital role in the fast growth of the fish which increases the protein content in the flesh. The components of feed are deoiled rice bran, groundnut oil cake, cotton seed cake and soyabean, out of these deoiled rice bran applications is more. Oilcakes in the feed are usually increased during the last 2 to 3 months prior to harvesting, in order to achieve fattening. The feed is given through bags suspended in the pond water at different places from bamboo poles. The bags are provided with holes through which the feed falls into water. Boats made of tin are used to carry the feedbags inside the pond. Feed is given once in a day usually in the morning time. The quantum of supplementary feed so given increases progressively during the culture period proportionately to the increase in the fish biomass. It is estimated that cost of supplementary feed accounts for 70% of the total variable cost.

DISEASE OUTBREAK AND CONTROL

Due to intensive use of fertilizers and feed disease occurrence is observed. Fungal, viral disease are observed which are treated by applying various medicines namely nuwan, butox, haldi, visaka *etc*. These medicines are properly diluted with water and spread throughout the water body.

HARVESTING AND MARKETING

Generally harvesting is done at the end of 10-12 months. Very few farmers are keeping the stock upto 15 months. Harvesting also depends upon the market price and demand. Harvesting is done by the middleman in which wage for labours and net charges are paid by the contractor. Near the pond site cost of fish varies in between Rs. 20-30 per kilogram, depending upon size of the fish. Initial packing is done at the pond site and final packing is done at the ice plant. Plastic crates are used for packing fish, in which 20 kg fish and 20 kg ice are kept in different layers. It is well known that fish is a perishable commodity which gets rotten within few hours. As Kolleru product is transported to longer distances of at least one thousand Kilometers it needs very meticulous methodology of packing which will keep it fresh for hours. Rice husk stored at the rice mills are also used by the packers to slowdown the melting process of ice. Production varies from 10-14 t/ha. Produce of the Kolleru Lake distributed to entire Eastern India and Bangladesh also. Daily 100-150 lorries from Kolleru Lake are flagged off to different parts of India.

ANCILLARY INDUSTRIES

Due to the growth of aquaculture, ancillary industries like ice factories, feed mills, lorry supply agencies, packing centers, and fish need centers and boat-making centers are also developed. Many ice plants have been established in the villages namely Akivvedu, Kaikaluru, Sreeparru and others.

SOCIO ECONOMIC CONDITIONS

Socio economic conditions of the fish farmers in Kolleru Lake area have undergone a significant change. The level of income increased manifold due to the existence of aquaculture in large scale. Many of the farmers found to have mobile phones, two and four wheelers in addition to modern electronic gadgets.

Kolleru Lake area is flourishing in aquaculture industry resulting in boosting of economy. The outcome of huge socio-economic information pertaining to aquaculture lies in proper analysis and interpretation of the obtained data from the field survey. The CD may be the considered as the weapon in the hands of researchers involved in studying the socio-economic aspects of aquaculture industry in Kolleru lake area. The information furnished in the database may be the guideline for future researchers and immensely helpful to all scientific and farming community all over India.

CREATION OF SOCIO ECONOMIC DATABASE OF CARP CULTURE IN KOLLERU LAKE

The Database has been created at CIFA covering many aspects of farmers associated with related socio-economic and farm specific variables. This would help in in-depth details of socioeconomic conditions of fish farmers and easy retrievals of the required data for analysis for unearthing of many hidden fact (Roy *et al.*, 2004)

Broad areas of coverage in this database system covers are:

- Fish farmers personal details
- Pond details
- Assets required
- Loans available
- Assistant provided
- Insurance offer
- Types of species used for culture
- Feeds, chemicals, fertilizers used
- Diseases incurred
- Drugs used
- Trainings
- Technologies used etc.

To store the above data, database is required which is the collection of related data stored together without unnecessary redundancy to serve multiple applications under the control of a database management system (DBMS).

Database provides the following features:

- Provides easy assess to the users for entering information through the forms
- Facilitates the user to query and get the right information as and when required
- Provides valuable information in the form of reports
- The system avoids the duplication of the same information

In this present database, three forty-five parameters are identified and distributed in sixty-six tables which store the large information of socio-economic data. Since records in the table contain the several categories of information, a field is added to each category to store the data in this database. The sixty-six tables, which bear the following heads along with corresponding field names, are shown in table 1 and the data types used in the present database are shown in table 2.

Table 1: Shows detailed heads and field name.

Heads	Field name
Farmer Details	Place of study, Case No., Date of interview, Reference year, Farmer name, Age, Education, Religion, Caste, Number of children, Village, District, State, Ex. Ch. Fish culture
Asset master	Asset code, Asset name
Asset	Place of study, Case No. Asset code, No. of items, Purchase value, Year of purchase
Assistance	Place of study, Case No. Assistance code, Name of Programme assisted, Year, Type of assistance
Assistance Master	Assistance code, Assistance agency, No. of farmer
Chemical master	Chemical code, Chemical name, Chemical rate
Chemicals applied	Place of study, Case no., Chemical code, Chemical name, Purpose, Quantity, Total cost
Disease Control	Place of study, Case no. Disease code, Disease name
Disease master	Disease code, Disease name
Disease master	Place of study, Case no. Disease code, Disease name
Drug master	Drug code, Drug name, Drug bie
Drugs Applied	Place of study, Case no., Drug code, Quantity of drugs on liter or kg per ac, Rate of Rs. Per kg or L, Method of application, Advice taken, Total cost
Feed Application	Place of study, Case no, Feed code, Rate of application, Rate (Rs./kg), Method of application, Time of application, Frequency per month, Total cost
Feed master	Feed code, Feed name, Feed rate
Fertilizer application	Place of study, Case no, Fertilizer code, Type of fertilizer, Rate of application, Total cost
Fertilizer master	Fertilizer code, Fertilizer name
Harvesting	Case no., Frequency of harvesting, Time, Labour employment,

(Contd.)

Heads	Field name
	maydays, Wage rate per maydays, Labor cost, Number of net, Rate per net, Net cost, Total harvesting cost
Income master	Source code, Name
Insurance	Place of study, Case no, Inscode, Purpose, Lyear, amount
Labour employed	Place of study, Case no, Labour code, Labour type, No. of labours, No. of days, Total cost
Labour master	Lcode, Ltype, Rate
Lime Application	Case no, Frequency of crop period, Time, Quantity in kg per ac, Rate in Rs. Per kg
Loan	Case no. Loan code, Sourcing agency, Amount, Lyear, Unpaid amount
Loan master	Loan code, Loan type
Machine master	Machine code, Machine type, Rate per hour
Netting Detail	Place of study, Case no, Purpose, Frequency per year, Ownership of netting, Labour and net cost per mandays
Pond Detail	Place of study, Case no, Ownership type, Lease value, Size of pond, No. of ponds, Depth of water max, Depth of water min, Colour of water, Cost of construction, Cost of renovation
Production Detail	Place of study, Casenor, Species code, Maximum weight (in kg), Minimum weight (in kg), Percentage of survival, Duration of seed (in day), Quantity of production (ton/ac), Selling price (Rs/kg), Total cost (in Rs.)
Production of species	Place of study, Case no, Species code, Species name, Quantity of production, Selling price, Total cost
Advice taken of application of drugs	Code, Advice taken from, No. of farmer, percentage
Attitude of farmers	Code, Attitude of farmers, No of farmers, percentage
Calamities faced by farmers	Code, No. of farmer, percentage
Caste	Caste code, Caste, No of farmers, percentage
Disease master	Sl. No., Disease code, Disease name, No of farmer, percentage
Farmer education	Education code, Education level, No of farmer, percentage
Fertilizer type frequency	Code, Type of fertilizer, No of farmer, Rate of application, Cost range
Frequency of application	Code, Frequency of application, No of farmer, percentage
Harvesting	Frequency of harvesting, No of farmer
Labour master	Lcode, Ltype, Rate, No of farmer, Percentage
Lime application	Sl No, Frequency, No of farmer, Percentage, Quantity
Loan	Source, No of farmer, Amount
Method of application of drugs	Code, Method of application, No of farmer, Percentage
Method of application	Method of application, No of farmer, Percentage
Netting detail	Code, Purpose of netting, No of farmer, Percentage
No of children	Children code, Children, No of farmer, Percentage
Pond detail ownership Type	Code, Ownership type, No of farmer, Percentage
Pond master	Pond code, Pond size, Size of pond, Size of Pondl, Size of pondh, No of farmer, Percentage

(Contd.)

Heads	Field name
Pond master I	Pond code, Size of pond1, Size of pond, Size of pondl, Size of pondh, No of farmer, Percentage
Rearing of fry	Sl No, Wherrearingisdoneornot, No of farmer, Percentage, Density
Religion	Religion code, Religion, No of farmer, Percentage
Risk attitude_fish culture profitability	Code, Fish culture profitability, No of farmer, Percentage
Risk attitude_increase in production	Code, Increase in production, No. of farmer, Percentage
Stocking Size	Code, Stocking size, No of farmer, Percentage, Stocking density
Stocking Type	Code, Stocking type, No of farmer, Stocking density
Time of application	Code, Time of application, No of farmer, Percentage
Type of integration	Code, Type of integration, No of farmer, Percentage
Water management	Code, Source of water, No of farmer, Percentage
Rearing of fry	Place of study, Case no, Species code, Whether rearing is done, Rearing density (no. per ac), Area of rearing pond (in ac), Size harvested (in gm), Rearing cost (in Rs per ac), Duration of rearing (in month)
Risk attitude	Place of study, Case no, Fish culture profitability, Increase in Pond, No. of calamities, Attitude of farmer
Source of income	Place of study, Case no, Source code, Source name
Species detail	Place of study, Case no, Species code, Species name, Maximum weight, Minimum weight, Percentage of survival, Duration of seed
Species master	Species code, Species name, Duration range, Range of weight, Quantity range, Selling price range
Stocking detail	Place of study, Case no, Species code, Seed type, Seed size (in gm), Lower seed size, Upper seed size, Source of seed, Amount of seed, Rate of seed
Technology	Place of study, Case no, Type of technology, Type of integration
Training	Place of study, Call no, Name of training, Stipend obtained
Water management	Place of study, Case no, Machine code, Source of water, Water cess in Rs/ac, Total cost, Total diesel consumed in life, Diesel rate in Rs/litre, Total diesel cost, Grand total water cost, Pond size in ac

Table 2: Shows data types, stores and size.

Data type	Stores	Size
Text	Alphanumeric characters	Up to 255 bytes
Number	Numeric values (integers or fractional values)	1, 2, 4 or 8 bytes
Date/time	Dates and time	8 bytes

Field properties like field size, format, and decimal places are used to prevent incorrect data entry. To increase efficiency, a primary key was included for case no., filled in each table not to allow duplicate or null values. And this will also help in easy association of the data from all the 66 tables. As because information is divided into different tables, relational database management system of one to many relationship between the tables are created.

SYSTEM DESIGN AND DEVELOPMENT

Back-end

Taking Visual Basic 6.0 as front-end and MS Access as back-end, this database system was developed, 66 tables are made to contain data used for storage and retrieval by the system. These tables are being linked together by different relations among themselves through ER diagram.

FRONT-END

The front-end is designed in the powerful programming language Visual Basic 6.0. There is one MDI form for the main menu screen. And several other forms to support the data entry and retrieval of data from the database. There are several data reports showing the summary data in organized way and also there are facilities for the user to see the query. The MDI form contains several menus and sub-menus. The first two menus are Master and Transaction, which are used to enter data into the system. Towards the bottom of each form there will be several command bottoms having a matching name to work accordingly.

The next menus are Query, Acknowledgement and Exist. Queries are used for retrieving information from the database and shows to user in an organized way through reports. There is also facility to take print of the same in case hard copy is required.

FISH DISEASE DATABASE IN CD-ROM:

This CD-ROM contains Fish Disease Database Information system for aquaculture farms. The system covers structure, design and its implementation in user defined research programs and to record and process disease related events in the farms. It includes standardized definitions of disease events, systematic classification of diseases, standard disease indices for recording and processing epidemiological data. There are 400 pages, which include 250 photographs depicting fish disease events (Sahu *et al.*, 2001).

For users: Autorun facility

SYSTEM REQUIREMENT

Microsoft windows
Pentium or later windows 95 or higher
16 MB RAM
VGA monitor

REFERENCES

Roy, A. K., 2004. CD-ROM Databases, Software's and List servers related to aquaculture research. *In*: Bioinformatics and Statistics Vol.I (Eds. Roy, A. K., A. Antony, M. Rout). Bioinformatics Centre, Central Institute of Freshwater Aquaculture (CIFA), Kausalyaganga, Bhubaneswar. pp. 119-123.

Saradhi, K. P., Roy, A. K., Sao, K.N. and Saha, G. S. 2002. Socioeconomic Database Management in Aquaculture. In: The Fifth Indian Fisheries Forum Proceedings (Eds. S. Ayyappan, J. K. Jena and M. Mohan Joseph), APSIB, Mangalore and AoA, Bhubaneswar, India.

Sahu, B.B. *et al.*, 2001. Fish disease database information system IVLP-NATP, CIFA Centre and Bioinformatics Centre on Aquaculture, CIFA, Kausalyaganga.

22

Proteomics and Bioinformatics

A. Subash Chandira Bose, R. Venkateshwaran and M. Balakrishnan

The dream of having genomes completely sequenced is now a reality. The complete sequence of many genomes including the human one is known. However, the understanding of probably half a million human proteins encoded by less than 30'000 genes is still a long way away and the hard work to unravel the complexity of biological systems is yet to come. A new fundamental concept called proteome (PROTEin complement to a genOME) has recently emerged that should drastically help to unravel biochemical and physiological mechanisms of complex multivariate diseases at the functional molecular level. Proteomics involves the systematic study of proteins in order to provide a comprehensive view of the structure, function and regulation of biological systems. Coupled with advances in bioinformatics, this approach to comprehensively describing biological systems will undoubtedly have a major impact on our understanding of the phenotypes of both normal and diseased cells. Initially, proteomics focused on the generation of protein maps using two-dimensional polyacrylamide gel electrophoresis.

INTRODUCTION TO PROTEOMICS

There's little doubt that proteomics — the study of an organism's complete complement of proteins — will have great impact in all areas of the life sciences in the years to come. And the reason is clear. "To really understand biological processes, we need to understand how proteins function in and around cells since they are the functioning units," says Hanno Steen, director of the Proteomics Center Boston.

The task of studying the proteome has its share of challenges. One involves the sheer number of proteins that need to be identified. The 35,000 genes in the human genome can code for at least ten times as many proteins; in extreme cases a single gene alone can code for over 1,000. Another challenge is that amino acids — the base units of proteins — are so small. Each amino acid is made from anywhere between 7 and 24 atoms. This is far beyond the reach of even the most powerful microscopes. How are researchers able to determine the sequence of amino acids that make up proteins? One way is by separating the proteins, breaking them up into smaller pieces, and using mass spectrometers to, in effect, "weigh" each amino acid. Each type of amino acid has a unique mass, making identification relatively straightforward. By identifying and sequencing these smaller pieces, researchers can then determine the identity of the protein they make up.That's obviously a simple explanation of the process. For a more in-depth explanation, check out our Guide to Sequencing and Identifying Proteins.

PROTEIN BASICS

A protein is a large, organic molecule made up of smaller molecules of amino acids. Although small proteins can contain as few as 50 amino acids, they are typically much larger, consisting of hundreds or even thousands of amino acids. Proteins are sometimes intentionally cut into smaller segments. These segments are called peptides. A peptide can consist of anywhere between two and 50 amino acids.

In a protein, amino acids are linked one after another, forming a long chain. These are held together by peptide bonds, which are chemical bonds between the carbon and nitrogen atoms of the amino acids. In addition to carbon and nitrogen, amino acids contain hydrogen, oxygen and sometimes sulfur.

Proteins contain 20 types of amino acids. With the exception of leucine and isoleucine, which share the same atomic mass of 113 Daltons, each amino acid has a unique mass. (Dalton is the unit of measurement for atomic mass. One Dalton is equal to 1/12th the mass of one atom of carbon 12.)

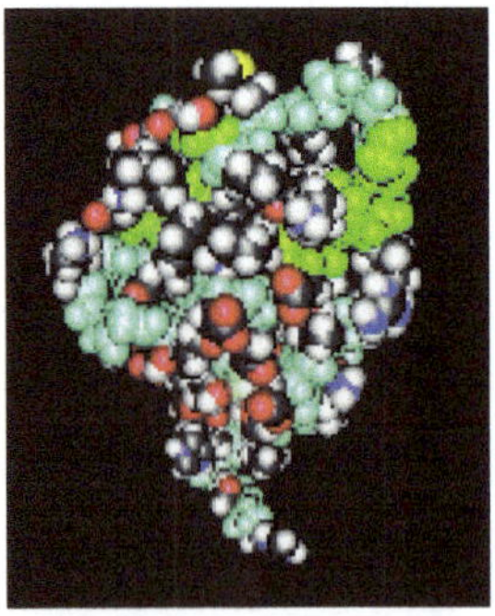

Fig.1: A single, long chain of amino acids gives this relatively small protein its three-dimensional shape.

WHAT DO PROTEINS DO?

Making up 12 to 18 percent of a lean, adult body, proteins play many crucial roles in human physiology. Here are a few:

- Proteins form the structural framework of tissues and other structures in the body. For example, collagen helps make up tendons, ligaments, cartilage and other connective tissue.
- Proteins carry substances throughout the body. For instance, hemoglobin is a carrier of oxygen and carbon dioxide.
- Proteins work as enzymes to facilitate chemical reactions. There are over 1,000 known types of enzymes in the human body.
- The proteins myosin and actin, allow muscles to contract.
- Proteins regulate physiological processes and control growth.
- Proteins are an integral part of the immune system. Antibodies, for example, consist of a type of protein used by the immune system.

HOW DO GENES CODE FOR PROTEINS?

You can think of the human genome as one long string of DNA that is separated into 46 pieces called chromosomes. Within each chromosome are thousands of genes. The DNA sequence that makes up a gene is like software code that instructs the cell how to make proteins. Four types of bases make up the DNA strand — adenine (A), cytosine (C), guanine (G) and thymine (T). Each base has a complement — another base that it usually pairs off with. A's pair off with T's, and C's pair off with G's. In all, the human genome is 3 billion base pairs long.

Three successive bases in the DNA strand, called a codon, can act as an instruction for the cell to select a specific type of amino acid. A series of codons instruct the cell to piece together a string of amino acids to form a protein. The cell knows when to begin and stop coding for proteins by recognizing special start and stop codons that are within the DNA strand.Genetic instructions are carried out by various molecules and cell structures, including transfer RNA (tRNA), messenger RNA (mRNA) and ribosomes. In other words, the cell is able to read pieces of DNA code and use the information to put a protein together. Amino acids come in an assortment of shapes. The cell, instructed by genes, takes advantage of this property and builds proteins that have vastly different sizes and shapes.

PROTOCOLS

1. Protein Extraction
2. Capillary Electrophoresis
3. Protein Digestion

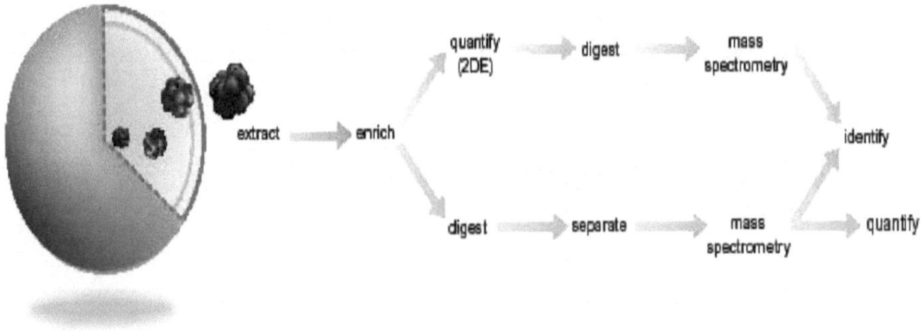

Fig 2. Protein extraction from whole tissues SDS extraction followed by acetone precipitation - simple extraction protocol that does not require phenol.

RECOMMENDED START PROTOCOL FOR WHOLE TISSUE EXTRACTIONS

1. Grind 1 g of fresh tissue to a powder with liquid nitrogen in a mortar and pestle.
2. Add 5 mL of extraction media (0.175 M Tris-HCl, pH 8.8, 5% SDS, 15% glycerol, 0.3 M DTT) directly to mortar and continue grinding for an additional 30 sec.
3. Filter homogenate through two layers of miracloth into a 50 mL Falcon tube at room temperature.
4. Immediately add 4 volumes of ice cold 100% acetone to filtered homogenate, mix by vortexing and place at -20°C for at least one hour to precipitate proteins.
5. Centrifuge at 5000 g for 15 min to collect precipitated protein, decant supernatant.
6. Gently blot residual acetone from container with Kimwipe and then wash pellet in 15-20 mL of cold 80% acetone. Be sure to thoroughly break-up pellet by pipetting, vortexing or sonication.
7. Repeat steps 5 and 6.
8. Collect final protein precipitate by centrifugation at 5000 g for 15 min and dry pellet by inverting on Kimwipe for 15 min at 37°C.
9. Resuspend final pellet in 0.5-1 mL of IEF extraction solution (8 M urea, 2 M thiourea, 2% CHAPS, 2% Triton X-100, 50 mM DTT, 2mM tributylphosphine) by pipetting and vortexing at 25-30°C. Incubate sample for 1 h at room temperature with agitation. Do not heat sample under any circumstances as this will lead to carbamylation of proteins.

10. Centrifuge for 10 min at 12000 g and use supernatant to rehydrate IPG strips.

11. If protein quantitation is necessary, precipitate protein sample with TCA or acetone prior to performing Bradford or Lowry assay as detergents and reducing agents interfere with these assays. Alternatively, use the new protein quantitation kits (e.g. EZQ from Invitrogen) that tolerate detergents and reducing agents.

CONCLUSION

Proteomics is a innovative technique to detect the protein structure and is often considered the next step in the study of biological systems, after genomics. Scientists are very interested in proteomics because it gives a much better understanding of an organism than genomics. First, the level of transcription of a gene gives only a rough estimate of its level of expression into a protein. This paper makes an attempt to put forward variuos application and future prospects of this field.

REFERENCE

Arora, Pankaj, S., *et al.* 2005. "Comparative evaluation of two two-dimensional gel electrophoresis image analysis software applications using synovial fluids from patients with joint disease". *Journal of Orthopaedic Science,* 10(2):160-166.

Liebler, D.C. 2002. Introduction to proteomics: tools for the new biology. Humana Press, Totowa, NJ. ISBN 0-585-41879-9 (electronic, on Netlibrary?)

Twyman, R.M. 2004. Principles of proteomics. BIOS Scientific Publishers, New York.

23

Data Mining and its Applications in Bioinformatics

Soma Mondal and M. Balakrishnan

Data Mining is the process of automatic discovery of novel and understandable models and patterns from large amounts of data. Bioinformatics is the science of storing, analyzing and utilizing information from biological data such as sequences, molecules, gene expressions, and pathways. Development of novel data mining methods will play a fundamental role in understanding these rapidly expanding sources of biological data. Data mining techniques are an automated means of reducing the complexity of data in large bioinformatics databases and of discovering meaningful and useful patterns and relationships in data. Bioinformatics, the application of computational techniques to analyze the information associated with biomolecules on a large-scale, has now firmly established itself as a discipline in molecular biology, and encompasses a wide range of subject areas from structural biology, genomics to gene expression studies. It provides an introduction and overview of the current state of the field. The main principles that underpin bioinformatics analyses look at the Data Mining in Bioinformatics and databases that are commonly used, and finally examine some of the studies that are being conducted, particularly with reference to transcription regulatory systems.

INTRODUCTION

Bioinformaticsis the recording, annotation, storage, analysis, and searching/retrieval of nucleic acid sequence (genes and RNAs), protein sequence and structural information. This includes databases of the sequences and structural information as well methods to access, search, visualize and retrieve the information. It is the process by which testable hypotheses are generated regarding the function or structure of a gene or protein of interest by identifying similar sequences in better characterized organisms.

Bioinformatics, the study of how information is represented and transmitted in biological systems, is a data-intensive field of research and development. It encompasses networking, databases, visualization techniques, search engine design, statistical techniques, modeling and simulation, AI and related pattern matching, and (the subject of this article) data mining. In bioinformatics, data mining is concerned with discovering how simple base pairs can be combined in different ways, many of which are unknown, to provide the form and function of the larger building blocks of life. Mastering this Biological data including discovering many of the underlying rules, relationships, and meanings requires human intelligence and intuition, leveraged by computer based tools.

The aim of this article is to introduce data mining techniques as an automated means of reducing the complexity of data in large bioinformatics databases and of discovering meaningful, useful patterns and relationships in data. The following sections provide an overview of the methods, technologies, and challenges associated with data mining.

KNOWLEDGE DISCOVERY

Data mining is one stage in an overall knowledge discovery process. This process involves selection and sampling of the appropriate data from the database(s); preprocessing and cleaning of the data to remove redundancies, errors, and conflicts; transforming and reducing data to a format more suitable for the data mining; data mining; evaluation of the mined data; and visualization of the evaluation results.

SELECTION AND SAMPLING

Data Mining isn't simply about searching for every possible relationship in a database. In a large database or data warehouse, there may be hundreds or thousands of valueless relationships. Because there may be millions of records involved and thousands of variables, initial data mining is typically restricted to computationally tenable samples of the holding in an entire data warehouse. The evaluation of the relationships that are revealed in these samples can be used to determine which relationships in the data should be mined further using the complete data warehouse. With large complex databases, even with sampling, the computational resource requirements associated with non-directed data mining may be excessive.

PREPROCESSING AND CLEANING

The bulk of work associated with knowledge discovery is in preparing the data for the actual analysis associated with data mining. The major preparatory activities include the following:

- Data Characterization—creating a high-level description of the nature and the content of the data to be mined.
- Consistency Analysis—determining the statistical variability in the data, independent of the domain.
- Domain Analysis—validating the data values in the larger context of the biology.
- Data Enrichment—drawing from multiple data sources to minimize the limitations of a single data source.
- Frequency and Distribution Analysis—weighing values as a function of their frequency of occurrence.
- Normalization—transforming data values from one representation to another.
- Missing Value Analysis—detecting, characterizing, and dealing with missing data values.

TRANSFORMATION AND REDUCTION

In the transformation and reduction phase of the knowledge discovery process, data sets are reduced to the minimum size possible through sampling or summary statistics. For example, tables of data may be replaced by descriptive statistics such as mean and standard deviation.

DATA MINING METHODS

The process of data mining is concerned with extracting patterns from the data by using techniques such as classification, regression, link analysis, segmentation, or deviation detection. Classification involves mapping data into one of several predefined or newly discovered classes. Regression methods involve assigning data a continuous numerical variable based on statistical methods. One goal in using regression methods is to extrapolate trends from a few samples of the data. Link analysis involves evaluating apparent connections or links between data in the database. Deviation detection identifies data values that are outside of the norm, as defined by existing models or by evaluating the ordering of observations. Segmentation identifies classes or groups of data that behave similarly, according to an established metric. These methods of data mining are typically used in combination with each other, either in parallel or as part of a sequential operation.

EVALUATION

In the evaluation phase of knowledge discovery, the patterns identified by the data mining analysis are interpreted. Typical evaluation ranges from simple statistical analysis and complex numerical analysis of sequences and structures to determining the clinical relevance of the findings.

VISUALIZATION

Visualization of evaluation results can range from simple pie charts to 3-D virtual reality displays that can be manipulated by haptic (force feedback) controllers.

CONCLUSION

With the current data, computational methods have become indispensable to biological investigations. Originally developed for the analysis of biological sequences, bioinformatics now encompasses a wide range of subject areas including structural biology, genomics and gene expression studies. Here, it provided an introduction and overview of the current state of field. In particular, we discussed the data mining techniques and databases that are commonly used, examined some of the studies that are being conducted with reference to Data Mining and finally looked at several practical applications of the field.

REFERENCES

Bergeron B. Applied for Bioinformatics Computing: Data Mining.

Frank, E., Hall, M., Trigg, L., Holmes, G. and Ian H. Witten Data Mining in Bioinformatics.

Luscombe, N.M., Greenbaum, D. and Gerstein, M. What is Bioinformatics? An Introduction and Overview.

24

Molecular Phylogenic Location of the Indian Liver Fluke, *Fasciola* (Trematoda: Fasciolidae) Based on The Ribosomal Internal Transcribed Spacer Regions

P. K. Prasad, V. Tandon, D.K. Biswal, L.M. Goswami and A. Chatterjee

Abstract

The species of liver flukes of the genus Fasciola (Platyhelminthes: Digenea: Fasciolidae)are obligate parasitic trematodes residing in the large biliary ducts of the herbivorous mammalian hosts causing fascioliasis worldwide. While Fasciola hepatica has a cosmopolitan distribution, the other major species, i.e., F. gigantica is reportedly prevalent in the tropical and subtropical regions of Africa and Asia. To determine the phylogenic location of Fasciola sp. of Assam (India) origin based on rDNA molecular data,ribosomal ITS regions were sequenced (Accession numbers- EF027103 and EF027104) and compared with other species of trematodes in the family Fasciolidae that includes, Fasciolopsis buski and Fascioloides magna besides the species of Fasciola. NCBI databases were used for sequence homology analysis using BLAST; multiple sequence alignments were performed using ClustalW program.The phylogenetic trees constructed based upon the ITS (I & II) sequences by multiple tree building methods in MEGA revealed a close relationship with isolates of F. gigantica from China, Indonesia , Japan, Egypt and Zambia, the isolate from China with significant bootstrap values being the closest. Using the novel approach of molecular morphometrics that is based on ITS2 secondary structure homologies, phylogenetic relationships of the various isolates of fasciolid species have also been discussed. While comparing the ITS1 sequences through BLAST search the sequence of another Indian isolate designated as F. gigantica (Accession- EF198867) showed almost absolute match with F. hepatica. Hence on the basis of molecular similarity this particular isolate should be identified as F. hepatica and not F. gigantica. The results corroborate that the Fasciola species prevalent in the north eastern region of India is in fact Fasciola gigantica, and not F. hepatica, which otherwise is the most common liver fluke throughout the globe. Thus, the ITS sequences provide a reliable tool to be used as species markers. To ascertain intra-specific strain variations, if any, different geographical isolates of Fasciola sp from the region need to be studied. The results support a phylogenetic relationship between members of Fasciolidae, as suggested by morphological features. Further studies with additional molecular markers are needed to determine the population structure and divergence between the two closely related species of this genus.

BACKGROUND

The trematode flukes of the genus *Fasciola* (the sheep liver fluke) are parasites of herbivores and infect humans accidentally causing fascioliasis worldwide. The parasite is very cosmopolitan in distribution being found throughout all regions of the world, both temperate and tropical. *F. hepatica* is the causative agent of fascioliasis or 'liver rot' in ruminants, where it may be an important pathogen. Human infections with *F. hepatica* are found in areas where sheep and cattle are raised, and where humans consume raw watercress, including Europe, the Middle East and Asia [1]. However, a wide variety of other mammals including man may also be infected, ingestion of contaminated vegetation such as watercress being a common mode of infection. Infection with *F.gigantica* on the other hand is found more commonly in tropical regions of the world where it is quite cosmopolitan in distribution, areas affected including Africa, Asia, many Pacific islands including Hawaii (where it has been reported in man), the Middle East and Southern Europe and the south of the USA [2]. Both fasciolids follow a similar two-host life cycle, in which freshwater snails of the family Lymnaeidae act as intermediate or vector hosts and a broad spectrum of mammals, mainly herbivorous large-size species, act as definitive hosts, including humans [3].

In addition to the two major species of *Fasciola* described above, there have been a number of other species reported. These include *F. halli* (In cattle and sheep in the U.S.A), *F. californica* (In cattle and sheep in the U.S.A), *F. jacksoni* (In elephants in Africa and the Indian Subcontinent) and *F. nyanzae* (In the hippopotamus).

Identification of closely related species based on morphological characters alone can be difficult. This is particularly the case of soft-bodied animals such as digenean trematodes. The morphology of flukes found in biliary canals and gall bladder has been the subject of numerous studies and is well known [4]. However, recent advances in molecular biology, in particular the amplification of specific DNA regions via the polymerase chain reaction (PCR) and improved sequencing techniques, have been employed to resolve taxonomic issues related to various helminth parasites by comparing their DNA, utilizing genetic markers in nuclear ribosomal DNA (rDNA) and mitochondrial DNA in particular [5]. The rDNA, which codes for structural components of ribosomes, is particularly useful for genetic studies because it is highly repeated and contains variable regions flanked by more conserved regions [6]. The nuclear ribosomal DNA internal transcribed spacers (ITS1 and ITS2), which occur between the 18S, 5.8S and 28S coding regions, have proven useful for diagnostic purposes at the level of species [7-13]. *Fasciola* spp and isolates of *Fascioloides magna* from different geographical regions have been discriminated on the basis of ITS sequences [14]. The ITS2 sequences have also been used to characterize the liver flukes from mainland China, which include *F. hepatica*, *F. gigantica* and an intermediate genotype, including polymorphism among ITS2 copies within the same fluke individual [15]. The ITS1 has been used for only relatively few numbers of species [16], though the ITS1 studies confirms the results and conclusions previously reached with ITS2.

According to the molecular clock estimations based on ITS sequences, the relatively few ITS nucleotide differences between different regions of the world suggest a recent geographical diffusion of *F. hepatica* from Europe into the other continents [17]. It needs to be taken into account that *F. hepatica* and *F. gigantica* are evolutionarily relatively recent species having diverged about 19 million years ago, according to sequence analysis of cathepsin-L-like cysteine proteases [18].

Most phylogenetic studies using current methods have focused on primary DNA sequence information. However, RNA secondary structures are particularly useful in systematics because they include characteristics, not found in the primary sequence, that give 'morphological' information [19]. The novel approach of molecular morphometrics that is based on ITS2 secondary structure homologies is recently being used to study the phylogenetic relationships of various species. This method allows one to take into account the regions where multiple alignments are barely reliable because of a large number of insertions and deletions. For these reasons, molecular morphometrics appear to be complementary to classical primary sequence analysis in phylogenetic studies. The basic idea behind molecular morphometrics is to use the molecular structures as a direct source of measurable information. This method is based on the assumption that secondary structure can be phylogenetically as significant as primary sequence. In other words, one can consider the secondary-structure elements of the RNA molecule, i.e., the helices, loops, bulges, and separating single-stranded portions, as phylogenetic characters [20].

In the present study, our main objective was to describe the molecular phylogenic location of the Indian liver fluke on the basis of ITS sequences of the parasite collected from ruminant hosts. We also aimed at determining the species prevalent in the northeastern region of the country, by comparing the sequences generated by amplifying the ITS regions of the parasite DNA by primary sequence analysis and molecular morphometrics data.

MATERIALS & METHODS

PARASITE MATERIAL AND DNA ISOLATION

Adult *Fasciola* were obtained in live form from hepatic biliary ducts of freshly slaughtered cow, *Bos indicus*. The worms recovered from these hosts represented the geographical isolates from Assam and other parts of India and not from the mountainous regions of N.E India. Eggs were obtained from mature adult flukes by squeezing between two glass slides. For the purpose of DNA extraction, adult flukes collected from different host animals were processed singly; eggs recovered from each of these specimens were also processed separately. The adult flukes were first immersed in digestion extraction buffer (containing 1% SDS, 25 mg Proteinase K) at 37° C for overnight. DNA was then extracted from lysed individual worms by standard ethanol precipitation technique [21] and also extracted on FTA cards using Whatman's FTA Purification Reagent. The FTA cards were allowed to dry for 1 h at room temperature prior to punching. 2-3 sample discs of 1.2mm size were taken from the desired spot using coring device assuming a 25µl reaction and placed in PCR amplification tube. They were then washed with FTA Purification reagent and TE Buffer. The disc was allowed to dry at room temperature for 1h to make it ready for PCR. DNA from the eggs was extracted only with the FTA card technique.

DNA AMPLIFICATION, SEQUENCING AND ITS ANALYSIS

The rDNA region spanning the ITS regions was amplified from DNA obtained from the egg and adult stages of the fluke by PCR. We used the universal primers based on conserved ITS sequences of *Schistosoma* species following [22] as detailed below:

i) ITS2 region- 3S (forward): 5'GGTACCGGTGGATCACTCGGCTCGTG-3' and A28 (reverse): 5'-GGGATCCTGGTTAGTTTCTTTTCCTCCGC-3'

ii) ITS1 region-BD1 (forward): 5'GTCGTAACAAGGTTTCCGTA-3' and 4S (reverse): 5'TCTAGATGCGTTCGAA (G/A) TGTCGATG-3'

The PCR amplification was performed following the standard protocol [23] with minor modifications as described elsewhere [12,13]. The resultant PCR products were separated by electrophoresis through 1.6% (w/v) agarose gels in TAE buffer, stained with ethidium bromide, transilluminated under ultraviolet light and then photographed. The known size fragments of Phi X 174 DNA/ Hae III Digest in agarose gel were used as marker. For DNA sequencing, the PCR products were purified using Genei Quick PCR purification Kit, and sequenced in both directions using PCR primers on an automated sequencer by DNA sequencing services of TCGA, New Delhi and Bangalore Genei Ltd.

The DNA sequences were put to further analysis by using various Bioinformatics tools including similarity search BLAST (URL http://www.ncbi.nlm.nih.gov/blast) and phylogenetic prediction by CLUSTALW (URL http:// www.ebi.ac.uk/clustalw) for each query DNA sequence.

MOLECULAR PHYLOGENETIC ANALYSIS

Initially, the sequences were aligned using ClustalW multiple alignment [24] with the default gap and extension penalties used by this program. Phylogenetic tree-building methods presume particular evolutionary models. Therefore, while interpreting the results obtained, we considered different tree building models to entertain possible explanations. Only unique sequences were used in tree construction. ITS sequences were entered in the MEGA for construction of the phylogenetic trees using Maximum Parsimony and distance methods namely the Neighbor-Joining, UPGMA and Minimum Evolution. Branch support was given using 1000 bootstrap replicates in MEGA [25].

Phylogenetic analysis was also carried out using the Bayesian approach with combined datasets using Mrbayes 3.1 [26]. In this approach, each data partition is allowed to have a different evolution rate. In addition, each helix was partitioned as if it had evolved at a different rate. The evolutionary model for the sequence dataset was set to the

GTR model with gamma-distributed rate variation across sites and a proportion of invariable sites. Two separate chains were run with 10 million generations each to test for the convergence of the parameter estimates from two different departure points of the search.

PREDICTED ITS2 RNA SECONDARY STRUCTURES AND ANALYSES

Secondary structures of ITS2 sequences of various fasciolid species were reconstructed by aligning their sequences using Bioedit [27]. The acquired structures with restrictions and constrains were submitted in MFOLD [28]. RNA was folded at a fixed temperature of 37° C, and the structure choosen from different output files was the desired 6-helicoidal ring or the one with the highest negative free energy if various similar structures were obtained.

The manually obtained alignment with secondary structure information was used to construct a matrix for cladistic analysis based on molecular morphometrics using geometrical features and base numbering [29]. Thus, secondary structure helices were numbered and treated like characters that vary depending on their base number. The nucleotides were counted for each character and the number scaled as discrete character states ranging from 0 to 8 or by subtracting the lesser count from the rest (each nucleotide addition counted as a new character, 0 being the character with the lowest nucleotide number).

RESULTS

Phylogenetic Trees

Phylogenetic trees were obtained by comparing the sequences of *F. gigantica* and available ITS (I & II) sequences for other fasciolid species. Phylogenetic analyses using the various distance methods and character state method like Maximum

Parsimony show that the topology is similar among the trees obtained. Boot strapping of the sequences with Neighbour- Joining revealed significant support for the clade containing *F. hepatica*, *F. gigantica, F. buski*, *and Fascioloides magna*.The values of 70% and above in the bootstrap test of phylogenetic accuracy indicates reliable grouping among different members of fasciolids. The phylogenetic trees constructed based upon the ITS (I & II) sequences by multiple tree building methods in MEGA revealed a close relationship with isolates of *F. gigantica* from China, Indonesia, Japan, Egypt and Zambia, the isolate from China with significant bootstrap values being the closest. Phylogenetic tree clade analysis discriminated the *Fasciola* sequences into two separate clades, one for *F. hepatica* and other for *F. gigantica*. Phylogenetic trees constructed showed that the *F. gigantica* populations differentiated in Asian region giving several lineages. While one of them is mainly spread in South-Eastern Asia and the second lineage can be found in Central Asia. The latter might have migrated to Africa where the separation of two *Fasciola* species took place. While comparing the ITS1 sequences through BLAST search (ClustalW alignment), the sequence of another Indian isolate designated as *F. gigantica* (Accession- EF198867) from Indian Veterinary Research Institute, Bareilly showed almost absolute match with *F. hepatica*.

Secondary Stucture Analysis

Five predicted RNA secondary structures were reconstructed from the unique sequences to provide the basic information for phylogenetic analysis. The ITS2 nuclear region ranged from 720bp in *F. gigantica* India to a minimum length of 361bp in *F. gigantica* China. The topology based only on the predicted RNA secondary structure of the ITS2 region resolved most relationships among the species studied. Bayesian analysis of the alignment retained the same topology and supported the same branches as the primary sequences.

DISCUSSION

The taxonomy of *Fasciola* spphas been based mainly on morphological data complemented with ecological, cytological and pathological results as well as clinical manifestations. Morphological differences found on stained and mounted adult specimens have been widely used to discriminate between platyhelminth species [30]. It is possible to distinguish between adult *F. hepatica* and *F. gigantica* on the basis of morphology, but much variation exists. In places where hybrid populations occur, a range of morphologies that confound identification is seen. Differentiating between the two species is not possible on the basis of clinical, pathological, or immunological findings and their eggs are morphologically very similar [31]. Consequently, where both species occur concurrently or in overlapping geographical distribution, it is not possible to be certain as to which species is responsible for the disease. The low number of records of infection with *F. gigantica* may well be due to the lack of good tools to distinguish this species from *F. hepatica*[32].

The comparison of ITS sequences from worms of different hosts and of different countries indicates that there exists a high species-specific homogeneity. In the present study, primary sequence analysis revealed a close relationship between the query sequence (from NER India) and isolates of *F. gigantica* from China, Indonesia, Japan, Egypt and Zambia [14].Phylogenetictrees constructed showed that the groups of multiple closely related genotypes of *F. gigantica* from Asia are broadly sympatric and shallow geographical specialization of genotypes found for *F. hepatica* as well. Such a pattern is expected for species with high gene flow whose populations have not been sundered by long term biogeographic barriers [33]. *F. hepatica* could enter Europe from Africa through Central Asia as well as from South-Eastern Asia. In addition, the present populations of *F. gigantica* from Central Asia may be the results of natural (recent or ancient) hybridization between ancestral lineages of genotypes, which originated from Asian or/and African ancestors [34]. The sequence of another Indian isolate designated as *F. gigantica* (Accession- EF198867) from Indian Veterinary Research Institute, Bareilly showed almost absolute match with *F. hepatica*. Hence on the basis of molecular similarity this particular isolate should be identified as *F. hepatica* and not *F. gigantica*. These reports support a phylogenetic relationship between members of Fasciolidae, as suggested by morphological features.

On the other hand, secondary-structure analysis of the same data confirmed the results mentioned for primary sequence analysis. Nevertheless, because there were inconsistencies in the placement of a few *Fasciola* species, this study needs to be extended, in order to gain a better understanding of the systematic of this group as well as the evolution of their predicted ITS2 RNA secondary structures.

CONCLUSION

The results corroborate that the *Fasciola* species prevalent in the north eastern region of India is in fact *Fasciola gigantica*, and not *F. hepatica*, which otherwise is the most common liver fluke throughout the globe. Thus, the ITS sequences provide a reliable tool to be used as species markers. To ascertain intra-specific strain variations, if any, different geographical isolates of *Fasciola sp* from the region need to be studied. Lack of data on genotypic diversity of *Fasciola* species in Africa and India does not allow the origin of regional populations to be unambiguously determined. Further studies with additional molecular markers are needed to determine the population structure and divergence between the two closely related species of this genus.

REFERENCES

1. Mas-Coma, S., Esteban, J.G. and Bargues, M.D. 1999. Epidemiology of human fascioliasis: a review and proposed new classification. *Bulletin of the WHO*, 77: 340-346.
2. Hammond, J.A. 1974. Human infection with liver fluke *Fasciola gigantica. Trans R Soc Trop Med Hyg*, 68: 253-254.
3. Mas-Coma, S., Bargues, M.D. and Valero, M.A. 2005. Fascioliasis and other plant-borne trematode zoonoses. *Int J Parasitol,* 35: 1255-1278.
4. Mas-Coma, S. and Bargues, M.D. 1997. Human liver flukes: a review. *Res Rev Parasitol,* 57:145-218.
5. Blair, D., Campos, A., Cummings, M.P. and Laclette, J.P. 1996. Evolutionary biology of parasitic platyhelminths: the role of molecular phylogenetics. *Parasitol Today*, 12: 66-71.
6. Hillis, D.M. and Dixon, M.T. 1991. Ribosomal DNA: molecular evolution and phylogenetic inference. *The quarterly review of Biology*, 66: 411-453.
7. Morgan, J.A.T. and Blair, D. 1995. Nuclear rDNA ITS sequence variation in the trematode genus *Echinostoma*: an aid to establishing relationships within the 37-collar spine group. *Parasitology*, 111: 609-615.
8. Leon-Regagnon, V., Brooks, D.R. and Perez-Ponce, G. 1999. Differentiation of Mexican species of *Haematoloechus* Looss, 1899 (Digenea: Plagiorchiformes): Molecular and morphological evidence. *J. Parasitol*, 85: 935-946.
9. Tkach, V.V., Pawlowski, J. and Sharpilo, V.P. 2000. Molecular and morphological differentiation between species of the *Plagiorchis vespertilionis* group (Digenea, Plagiorchidae) occurring in European bats, with a re-description of *P. vespertilionis* group (Muller, 1780). *Syst Parasitol.*, 47: 9-22.
10. Kostadinova, A., Herniou, E.A., Barrett, J. and Littlewood, D.T. 2003. Phylogenetic relationships of *Echinostoma rudolphi,* 1809 (Digenea: Echinostomatidae) and related genera re-assessed via DNA and morphological analysis. *Syst Parasitol.*, 54:159-176.
11. Scholz, T., Skerikova, A., Shimazu, T. and Grygier, M.J. 2004. A taxonomic study of species of *Bothriocephalus* Rudolphi, 1808 (Cestoda: Pseudophyllidea) from eels in Japan: morphological and molecular evidence for the occurrence of *B. claviceps* (Goeze, 1782) and confirmation of the validity of *B. japonicus* Yamaguti, 1934. *Syst Parasitol.*, 57: 87-96.
12. Prasad, P.K., Tandon, V., Chatterjee, A. and Bandyopadhyay, S. 2007. PCR-based determination of internal transcribed spacer (ITS) regions of ribosomal DNA of giant intestinal fluke, *Fasciolopsis buski* (Lankester, 1857) Looss, 1899. *Parasitol Res.*, 101:1581-1587.

13. Tandon, V., Prasad, P.K., Chatterjee, A. and Bhutia, P.T. 2007. Surface fine topography and PCR-based determination of metacercaria of *Paragonimus* sp. from edible crabs in Arunachal Pradesh, Northeast India. *Parasito Res*., 102:21-28.
14. Huang, W.Y., He, B., Wang, C.R. and Zhu, X.Q. 2004. Characterisation of *Fasciola* species from Mainland China by ITS-2 ribosomal DNA sequence. *Vet Parasitol*, 120: 75-83.
15. Mas-Coma, S., Funatsu, I.R. and Bargues, M.D. 2001. *Fasciola hepatica* and lymnaeid snails occurring at very high altitude in South America. *Parasitology*, 123: S115-S1127.
16. Mas-Coma, S., Bargues, M.D., Valero, M.A., Fuentes, M.V. 2003. Adaptation capacities of Fasciola hepatica and their relationships with human fascioliasis: from below sea level up to the very high altitude. In: Combes C, Jourdane J, eds. Taxonomy, Ecology and Evolution of Metazoan Parasites, Vol.2. *Perpignan*: Perpignan University Press, 81-123.
17. Irving, J.A., Saithill, T.W., Pike, R.N., Whistock, J.C. and Smooker, P.M. 2003. The evolution of enzyme specificity in *Fasciola* spp. *J. Mol. Evol.*, 575: 1-15.
18. Caetano-Anolles, G. 2002. Tracing the evolution of RNA structure in ribosomes. *Nucleic Acids Res.*, 30: 2575-2587.
19. Zwieb, C., Glotz, C. and Brimacombe, R. 1981. Secondary structure comparisons between small subunit ribosomal RNA molecules from six different species. *Nucleic Acids Res.*, 9: 3621-3640.
20. Bowles, J., Blair, D. and McManus, D.P. 1995 A molecular phylogeny of the human schistosomes. *Mol Phylogenet Evol.*, 4:103-109.
21. Sambrooke, J., Fritsch, E.F. and Maniatis, T. 1989. Molecular cloning: a laboratory Manual, 2nd (ed). Cold Spring Harbor: Cold Spring Harbor Laboratory Press.
22. White, B.A. 1993. PCR Protocols, current methods and applications. Vol.15. Totowa, New Jersey: Humana press.
23. Thompson, J.D., Higgins, D.G. and Gibson, T.J. 1994. CLUSTAL W: improving the sensitivity of progressive multiple sequence alignment through sequence weighing, positions-specific gap penalties and weight matrix choice. *Nucleic Acids Res.*, 22: 4673-4680.
24. Hillis, D.M. 1993. Bull, J.J. An empirical test of bootstrapping as a method for assessing confidence in phylogenetic analysis. *Syst Biol.*, 42: 182-192.
25. Huelsenbeck, J.P. 2001. Ronquist, F. Bayesian inference of phylogenetic trees. *Bioinformatics,* 17: 754-755.
26. Billoud, B., Guerrucci, M.A., Masselot, M., Deutsch, J.S. Cirripede phylogeny using a novel approach: Molecular, Morphometrics. *Mol Biol Evol.*, 17:1435-1445.
27. Hall, T.A. 1999. BioEdit: a user-friendly biological sequence alignment editor and analsysis program for Windows 95/98/NT. *Nucl Acids Symp.*, 41: 95-98.
28. Zuker, M. 2000. Mfold web server for nucleic acid folding and hybridization prediction. *Nucl Acids Res.*, 31: 3406-3415.
29. Miyazaki I. 1974. Lung flukes in the world. Morphology and life history. *In*: A symposium on epidemiology of parasitic diseases, by Sasa M, ed.
30. Lotfy, W.M., Hillyer, G.V. 2003. *Fasciola* species in Egypt. *Experimental Pathology and Parasitology,* 11: 9-22.
31. Marcilla, A., Bargues, M.D., Mas-Coma, S. 2002. A PCR-RFLP assay for the distinction between *Fasciola hepatica* and *Fasciola gigantica*. *Mol Cell Probes.*, 16:327-333.
32. Avise, J.C. 2000. Phylogeography: The history and formation of species. Havard University Press, Cambridge, MA.
33. Itagaki, T., Tsutumi, K.I., Sakamoto, T., Tsutsumi, Y., Itagaki, H. 1995. Characterisation of genetic divergence among species within the genus *Fasciola* by PCR-SSCP. *Japanese Journal of Parasitology.* 44:233-246.

25

Prediction of Amino Acid Levels in Protein Sequences Using Artificial Neural Networks

K. Meena and M. Manimekalai

Abstract

A Mathematical Method for Artificial Neural Networks offers an original, broad and integrated approach that explains each tool in a manner that is independent of specific ANN systems. Artificial Neural Networks are abstract mathematical models of brain structure and functions. They are nothing but a mathematical model that relies on varying the strength of connections between the internal processing elements to interpret data. They can be applied to many pattern classification problems. In this paper, it has been used to predict the level of amino acid in protein sequences. Amino acids are the building blocks of proteins. It is any molecule that contains both amine and carboxylic acid functional groups. Here, known protein sequences of ingredients are taken as input patterns. The network is trained using back propagation algorithm with sigmoid function as activation function by generating random weights. It is tested again with unknown protein sequence. Error is calculated by the difference between actual network output and desired output. If the training is successful the patterns are correctly classified. The network is trained until the error is minimum. The amino acid level is thus predicted in a particular protein sequence.

INTRODUCTION

Amino acids are the molecular units that make up proteins. All proteins are various compositions of twenty specific naturally occurring amino acids. Two individual amino acids can be linked to form a larger molecule, with the loss of a water molecule as a by-product of the reaction. It is a textual representation of such a reaction. The newly created C-N bond between the two separate amino acid is called a peptide bond. The term peptide bond implies the existence of the peptide group that is commonly written in text as –COOH. Two molecules linked by a peptide bond become what is called a dipeptide[1]. A chain of molecules linked by peptide bonds is called a polypeptide. A protein is made up of one or more polypeptide chains, each of which consists of amino acids. Instead of writing out complex formule, sequences of amino acids are commonly written using the three – or one letter codes. Example ala-gly-val-leu-phe (3 letter) or AGVLF (1 letter).

TABULAR REPRESENTATION OF AMINO ACID

A table listing one- letter symbols, the three-letter symbols and the linear structure of the standard amino acids is given in *table1*

Table 1: List of Amino Acids

Name	Abbreviation	Linear Structure
Alanine	alaA	CH3-CH(NH2)-COOH
Arginine	argR	HN=C(NH2)-NH-(CH2)3-CH(NH2)-COOH
Asparagine	asn N	H2N-CO-CH2-CH(NH2)-COOH
Aspartic Acid	asp D	HOOC-CH2-CH(NH2)-COOH
Cysteine	cys C	HS-CH2-CH(NH)-COOH
Glutamic Acid	glu E	HOOC-(CH2)2-CH(NH2)-COOH
Glutamine	gln Q	H2N-CO(CH2)2-CH(NH2)-COOH
Glycine	gly G	NH2-CH2-COOH
Histidine	his H	NH-CH=N=CH=C-CH2-CH(NH2)-COOH
Isoleucine	ile I	CH3-CH2-CH(CH3)-CH(NH2)COOH
Leucine	ieuL	(CH3)2-CH-CH2-CH(NH2)-COOH
Lysine	lys K	H2N-(CH2)4-CH(NH2)-COOH
Methionine	met M	CH3-S-(CH2)2-CH(NH2)-COOH
Phenylalanine	phe F	Ph-CH2-CH(NH2)-COOH
Proline	pro P	NH-(CH2)3-CH-COOH
Serine	ser S	HO-CH2-CH(NH2)-COOH
Threonine	thr T	CH3-CH(OH)-CH(NH2)COOH
Tryptophan	trp W	Ph-NH-CH=C-CH2-CH(NH2)-COOH
Tyrosine	tyr Y	HO-Ph-CH2-CH(NH2)-COOH
Valine	val V	(CH3)2-CH-CH(NH2)-COOH

PROTEINS, OCCURENCE AND FUNCTIONS

The term protein is derived from the Greek word 'Protos' which means 'OF FIRST IMPORTANCE'. Proteins are found in all living things. They form essential constituents of protoplasm, cell membrane,or complex protein combined with lipids and nucleic acids. Genes, which carry hereditary characteristic or nucleic proteins. where they occur, they play vital role in the biological functions, and serve as buildings blocks for cellular and organic structure. Proteins form components of catalysts that are essential for the metabolic reactions. They are also constituents of most of the hormones, which are regulators of the metabolism of important biological components. Proteins form about 15% of the total body weight of the average adult. The protein in different tissues such as connective tissue,bones,muscles,brain,blood and other biological fluids differ in composition and properties. The protein content of plants is relatively less than that of animal. In the plants, seeds and cereals contain the highest protein content, while the leaves, stem and tubers contain much less protein. All protein contains carbon, hydrogen, oxygen and nitrogen. Some contain sulphur and some phosphours. An approximate composition of these elements would be carbon -15%,hydrogen-7%, oxygen -20%, nitrogen -60%, sulphur-1% and phosphorus -2%.It is assumed that all proteins contain about 16% nitrogen. The physiological significance and functions of proteins are varied but vital. They regulate and integrate the numerous physiological and metabolic processes in the body through harmones, enzymes and nucleoproteins [2]. The plasma proteins help in the maintenance of acid-base balance and osmotic pressure of the body fluids [3].

PROBLEM ANALYSIS

The aim is to predict amino acid level in protein sequences. An Artificial Neural Network with supervised algorithm like back propagation algorithm is used for predicting the amino acid level from the given protein sequence. Protein sequences of ingredients are taken for testing. Nonetheless, a large body of sequence information exists particularly in the form of expressed sequence tags (ESTs). The following sequence represents the protein sequences for some ingredients.

PROTEIN SEQUENCE FOR INGREDIENTS

MCRLQATHLL ARLFLLLALC TALVSSLTAV APPFGQLSVS GTNLVGANGQ PVQLIGNSLF
WHQWYPQFWN ADTVKALKCN WNANVIRGAM GVDEGGYLSD ANTAYNLMVA VIEAAISNGI
YVIVDWHAHN SHPDEAVKFF TRIAQAYGSY PHILYEDFNE PLSVSWTDVL VPYHKKVIAA
IRAIDKKNVI ILGTPTWSQD VDVASQNPIK DYQNLMYTLH FYASSHFTND LGAKLKTAVN
NGLPVFVTEY GTCEASGNGN LNTDSMSSWW TLLDSLKISY ANWAISDKSE ACSALSPGTT
AANVGVSSRW TSSGNMVASY YKKKSTGVSC SGSSSGSSSG SSSGSSGSSS GSSSGSSSGS SSGSSSGSSG SSSGSSSGSG SASISVVPSN TWNGGGQVNF EIKNIGSVPL CGVVFSVSLP

IMPLEMENTATION AND TESTING

The implementation of the work is explained through the following methods.

- Correlation and Regression Method
- Artificial Neural Network method

CORRELATION AND REGRESSION METHOD

Correlation and regression analysis has been made for the numerical representation data of protein sequence. The strength of the linear association between two variables is quantified by the correlation coefficient.

Given a set of observations $(x_1, y_1), (x_2, y_2)\ldots. (x_n, y_n)$, the formula for computing the correlation coefficient is given by,

$$R = \sum_{i=1}^{\gamma-1} \left(\frac{(x_i - \bar{x})}{s_x} \right) \left(\frac{(y_i - \bar{y})}{s_y} \right)$$

Where,

n= Total Number of values

x_i=No. of motifs in the known protein sequences.

y_i=No. of motif in the unknown protein sequence

R=Correlation Coefficient

x=Mean value for x_i values.

Y =Mean value for y_i values

S_x=Standard deviation for x_i values.

S_y=Standard deviation for y_i values.

The correlation coefficient always takes a value between -1 and 1, with 1 or -1 indicating perfect correlation (all points would lie along a straight line in this case). A positive correlation indicates a positive association between the variables (increasing values in one variable correspond to increasing values in the other variable), while a negative correlation indicates a negative association between the variables (increasing values is one variable correspond to decreasing values in the other variable). A correlation value close to 0 indicates no association between the variables.

Since the formula for calculating the correlation coefficient standardizes the variables, changes in scale or units of measurement will not affect its value. For this reason, the correlation coefficient is often more useful than a graphical depiction in determining the strength of the association between two variables

By using this correlation and regression method, prediction of amino acid level in the above mentioned sequences the classification accuracy of 90% is achieved [4].

A) ARTIFICIAL NEURAL NETWORK METHOD

This consists of two modules:

1. Training
2. Testing

1) Training Module

(Training the Artificial Neural Network with known motifs along with target values)

In this module, known protein sequence is given input to the network. The protein sequence is the chain of 20 amino acids. The main program encodes the 20 amino acids and converts the characters in the protein sequence into the ASCII values. The ASCII values are normalized and then presented as an input pattern. A number of input motifs and their target values are presented to the network and weights between the layers are modified until the MSE error is reached [5]. This is illustrated in the following figure (Figure 1).

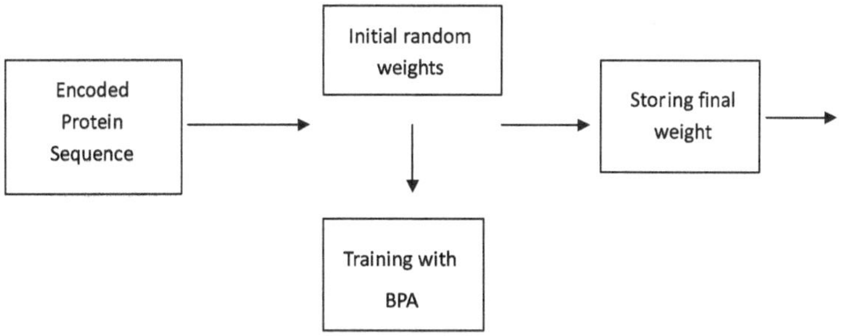

Fig 1. Training the network

2) Testing Module

(Testing the network using given sequence to find out the presence of motif)

A sequence is given to the network. If a motif is present, the level of the motif is displayed. This is possible by multiplying the given motif with the final or trained weights. This is illustrated in the above figure (Figure 2).

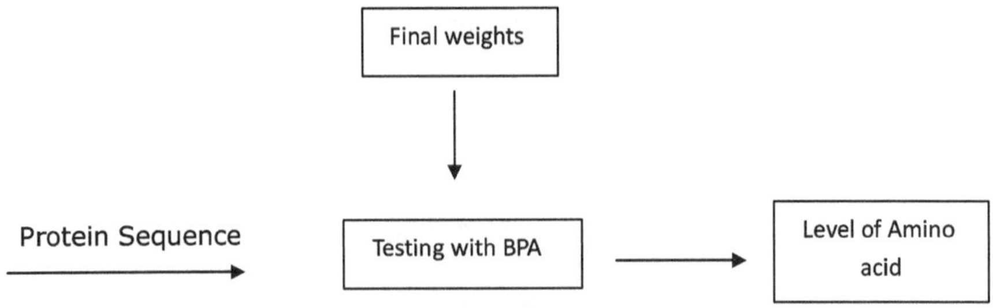

Fig 2. Testing the Network

The BPA uses the steepest-descent method to reach a global minimum. The number of layers and number of nodes in the hidden layers are decided. The connections between nodes are initialized with random weights. A pattern from the training set is presented in the input layer of the network and the error at the output layer is calculated. The error is propagated backwards towards the input layer and the weights are updated. This procedure is repeated for all the training patterns. At the end of the iteration, test patterns are presented to ANN and the classification performance of ANN is evaluated. Further training of ANN is continued until the desired classification performance is reached [6].

RESULT ANALYSIS

The size of the network used for training the protein sequence is 15*10*1. Initial weights are chosen from 0.25 to 0.55. The number of iterations taken for the network to reach the desired mean squred error of 0.001 is 11000[7]. It is found from the following table (Table 2) that the accuracy of the performance of the network is 96.5%.

Table 2:. A Table showing the presence of the motif

Protein Sequences	*Presence (P)/Absence (A)*
FLLKWTVILLLSVSF	P
WTRVYRLLKPSPNVV	P
AYSNITYYTRILPPV	A
MLRRNFRLDPQGTNL	P
GLGMYATLWLREHNR	P
PGLGMYATLWLREHN	P
QYQNRISVEFNQLJH	P
AGQIGGGHNIHPVVS	P
GLMGNPICSPDYWKP	P
VCLNTKWCPYVSFHT	P

During the training of the network, initial weights are given for the network and the final weights are stored into a file. During testing the final weights are used with the same configuration of the network and the protein sequence is presented to the network. If an amino acid sequence is present an indication is provided. A graph showing the value of Mean squared error and the number of iterations is

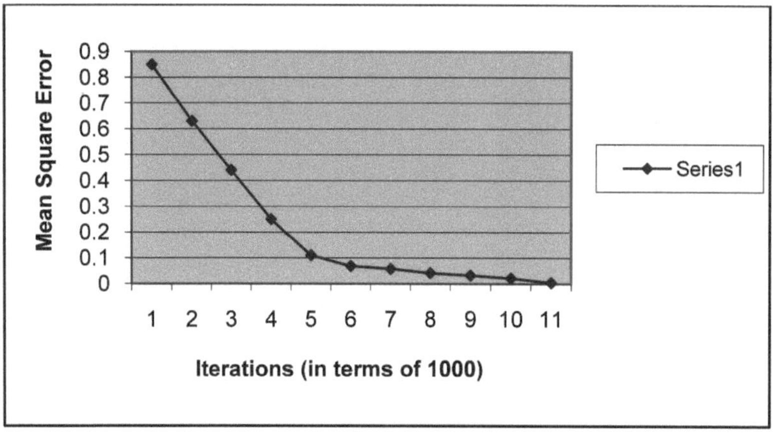

Fig 3: Graph showing the accuracy of prediction

given below.(Figure 3)

CONCLUSIONS

Soya bean has been taken as an example. The soyabean sequence is considered for finding the presence of modify like methionine etc., by using backpropagation algorithm with one input layer,one hidden layer and one output layer.The time taken for the neural network to reach the accuracy is minimum with mean squared error reaching to 0.0001 leading to 11000 iterations. During testing the accuracy of predicting the amino acid has resulted to 96.5%. It is found that artificial neural network performs better in the form of reduction in time, generalization capacity and be able to compensate if there is a corruption in the data.

REFERENCES

1. Gromiha, M.M., Selvaraj, S., Thangakani, A.M., Computational Biology Research Center (CBRC), National Institute of Advanced Industrial Science and Technology (AIST), Japan."A statistical method for predicting protein unfolding rates from amino acid sequence".
2. Jeffry, L. Shultz, Jeffry, D. Ray and David, A. Lightfoot, 2001. "A sequence based synteny map between soybean and Arabidopsis thaliana", BMC Genomics. 2007;8:8 Published online January 8.doi:10.1186/1471-2164-8-8.
3. Bock, J.C. Barker, B.C., Clinton, A.G, Wilson, M.B. and Lewis, F.R. University of California, San Francisco, "Post-traumatic changes in, and effect of colloid osmatic pressure on the distribution of body water".
4. Kristin M.Bedard, Sarah Daijogo and Bert L.Semler, "A nucleo-cytoplasmic SR protein functions in viral IRES-mediated translation initiation"

5. Thomas Nord, B., Jarvik, T., Bush, A.L., Shoemaker, R.C., Lark, K.G., Kahler, A.L., Kaya, N., Vantoai, T.T., Lohnes, D.G., Chung, J., 1999. "Cregan Prediction of protein secondary structure at 80% accuracy" *Crop Sci.*; 39:1464-1490.
6.Willam B. Roush and TERRi.L.Cravenar, Department of Poultry Sciences, Pennsylvannia University, Pennsylvannia,"Artificial Neural Network Prediction of Amino Acid Levels in feed ingredients".
7. Yamen Bai, John S. Milne, Leland Mayne, S.Walter Englander, The Johnson Research Foundation, Department of Biochemistry and Biophysics, University of Pennsylvania, Philadelphia, Pennsylvania, "Primary structure effects on peptide group hydrogen exchange".

26

Applying New Data Mining Method of Rough Sets to Predict Protein Structural Class

K. Meena and M. Durairaj

Abstract

The protein classification prediction is an important Problem in molecular biology, and one that has attracted a lot of attention, because there is a gap between sequence and structure of proteins from various resources. The proteins can be usually classified into one of the four structural classes; they are all-α, all-β, α/β and $\alpha+\beta$. To address the problem of protein classification, many different algorithms and efforts have been made so far. In this paper, a new rule based data mining method for the prediction of protein structural classes is constructed using Rough Sets algorithm. Rough Sets Theory (RST) is a new mathematical approach to study the realistic information (Z. Pawlak, 1982). It focuses on knowledge reasoning by the whole set directly approximating to the incomplete and uncertain information. There are studies on the knowledge reduction and features selection of mass data. To apply rough sets algorithm, a decision table is constructed using Amino acid compositions and eight physicochemical properties data as conditional attributes. The generated decision rules after reducing the decision system can be used to classify new objects.The classification results of Rough Sets is verified by using self-consistency and jackknife tests on the datasets constructed by G.P. Zhou and comparison is made with some of the previous works. The results showed that the rough sets approach is very promising *and may play a complementary role to the existing powerful approaches like Component-Coupled, Artificial Neural Network, Support Vector Machine (SVM), and LogitBoost.*

INTRODUCTION

The prediction of protein structural classes is put forwarded by Klein and Delisi, 1986 [1], and Klein brought discriminate analysis method to this problem [2]. Thousands of data of proteins can be analyzed on the high-throughput micro array platform in bioinformatics. There is a gap between sequence and structure; therefore, the prediction of protein structural classes is hot research field today. Choosing suitable theory to deal with protein data sets for structural classification is an interesting subject among researchers. Rough Sets is a very promising method in bioinformatics. Rough Sets Theory can be an attractive approach to extending or modifying incomplete knowledge base or domain theories. Usually a protein can be classified into one of the four structural classes: all-á, all-ß, á/ß and á+ß. There are many different algorithms and efforts have been made to address this protein structural classification. A systematic presentation and review about prediction of protein structural class and subcellular locations were introduced and compared with some existing methods by Chou [1]. In 1992, a weighting method [4] was proposed to predict protein structural classes from amino acid composition and a maximum component coefficient method was proposed by Zhang and Chou[5]. After that, an Artificial Neural Network based algorithm [6] was developed that considers six hydrophobic amino acid patterns together with amino acid compositions, and a cross-validation test was used to verify the accuracy of this method. A method based on the scale of Mahalanobis distance is proposed by Chou and Zhang [8] in 1994, and it incorporates the correlative effect among different amino acids automatically. The component-coupling algorithm proposed by Chou *et al.* [9] that took into account the coupling effect among different amino acid components.

This method was ever thought to be one of the most accurate algorithms to predict protein structural classes. In 2001, Cai *et al.* [11] introduced Support Vector Machine (SVM), a machine learning method based on statistical learning theory, to deal with this problem. Shen *et al.* [13] introduced supervised fuzzy clustering approach to this problem. Another recently developed powerful method, Logit Boost, was proposed by Fend *et al.* [14], which classifies protein through combining many weak classifiers together in order to build a stronger classifier.

In this study, a method based on supervised learning approach, Rough Sets Theory [15], to predict protein structural is proposed. In this method, two datasets constructed by Zhou [16] from SCOP were used to verify and test the efficiency of Rough Sets Theory. The Rosetta software developed by Øhrn [18], which is a freely available on internet for data mining with Rough Sets Theory, is used in this method for prediction of protein structural classification. Amino acid composition and 8 values of physicochemical properties were extracted from primary sequences of data-sets to construct the decision table.

MATERIALS AND METHODS

The self-consistency test is used to verify the performance of this Rough Sets Theory based method and cross-validation based on jackknife test for evaluating prediction results. These methods are considered the most rigorous and objective methods for evaluation of prediction with maximum stability. Self-consistency tests are performed against the two data-sets, and the results of self-consistency tests are showed in Table 1.

Table 1: Results of self-consistency test

Data Sets	277 domains				498 domains			
Algorithm	Component coupled	Artificial Neural Network	SVM	Rough Sets	Component coupled	Artificial Neural Network	SVM	Rough Sets
All-α	95.7%	98.6%	100%	100%	95.8%	100%	100%	100%
All-β	93.4%	93.4%	100%	100%	95.2%	98.4%	100%	100%
α/β	95.1%	96.3%	100%	100%	94.9%	96.3%	100%	100%
α+β	92.3%	84.6%	100%	100%	95.4%	84.5%	100%	100%
Overall rate of accuracy	94.2%	93.5%	100%	100%	95.8%	94.6%	100%	100%

All the percentages of correct prediction on both datasets reach 100%, which is the same as the results of Support Vector Machine method [11]. The results indicated that Rough Sets captured the characteristics between sequences and their classes through amino acid composition and physicochemical properties.

Table 2: Results of jackknife test

Data Sets	277 domains				498 domains			
Algorithm	Component coupled	Artificial Neural Network	SVM	Rough Sets	Component coupled	Artificial Neural Network	SVM	Rough Sets
All-α	84.3%	68.6%	74.3%	77.1%	93.5%	86.0%	88.8%	87.9%
All-β	82.0%	85.2%	82.0%	77.0%	88.9%	96.0%	95.2%	91.3%
α/β	81.5%	86.4%	87.7%	93.8%	90.4%	88.2%	96.3%	97.1%
α+β	67.7%	56.9%	72.3%	66.2%	84.5%	86.0%	91.5%	86.0%
Overall rate of accuracy	79.1%	74.7%	79.4%	79.4%	89.2%	89.2%	93.2%	90.8%

The results of Jackknife test performed on the datasets are shown in Table 2. Two sets of decision rules are generated as classifiers through the reduction of decision tables. The classifier trained by 277 domains contains 46651 decision rules in total, and the one of 498 domains contains 52474 decision rules. The distribution of 4 structural classes in decision rules is shown in Table 3. These rules can be used to classify new protein sequences to the 4 structural classes.

Table 3: Statistics of decision rules

	A	β	α/β	α+β	Total
277 domains	11711	10250	11886	12804	46651
Percentage	25.10%	21.97%	25.48%	27.45%	
	A	β	α/β	α+β	Total
498 domains	12744	11211	13771	14748	52474
Percentage	24.29%	21.36%	26.34%	28.11%	

ROUGH SETS THEORY

Rough sets theory, introduced by Pawlak [15] in the early 1980s, is a mathematical tool of representing and reasoning about imprecise and uncertain data. The rough set theory provides a formal framework for data mining. It has several favorite features such as representing knowledge in a clear mathematical manner, deriving rules only from facts present in data and reducing information systems to its simplest form, etc. Each 'if-then' rule induced from decision table identifies a minimal set of features discriminating one particular example from other classes. Rough sets theory distinguishes between objects based on the concept of *indiscernibility*, and deals with the approximation of sets by means of binary relations that is typically constructed from empirical data. The set of rules induced from all training examples constitutes a classificatory model capable of classifying new objects. The basic applications of rough set theory are: construction of decision table called *information system, approximations, reduction of attributes* and *others*.

INFORMATION SYSTEM

An information system (can also be called decision table) can be represented as A = (U, A) where ***U*** is a non-empty set of finite objects or instances called universe, the finite attribute set ***A*** is the union of condition '***C***' and decision attributes '***D***' (i.e. $A = C\ U\ D$). With every attribute a ∈ I we associate a set Va such that a: U → Va. The set Va is called the domain or value set of attribute a. With every subset *PQ* there is an associated equivalence relation $IND(P) = \{(x, y) \in U^2 \mid \forall q \in P, f(x,q) = f(y,q)\}$. If $(x,y) \in IND(P)$, then x and y are indiscernible by attributes from *P*. These indistinguishable sets of objects therefore define an equivalence relation, referred to as the *P*-indiscernibility relation. In this way, Rough Sets provide a simple form to treat with the uncertainty. Given information system I, let $X \subseteq U$ be a set of objects and $P \subseteq Q$ a selected set of attributes. The lower and upper approximations of X with respect to P are defined as: $P_*(X) = \cup\{Y \in U \mid IND(P) : Y \subseteq X\}$, $P^*(X) = \cup\{Y \in U \mid IND(P) : Y \cap X \neq \phi\}$. The P-boundary of X in the information system I, is defined as $BND(X) = P^*(X) - P_*(X)$. The tuple $(P_*(X), P^*(X)$ composed of lower and upper approximation is called a rough set. The accuracy of the rough set representation of the set X can be given [15] as: $\alpha_P(X) = card[P_*(X)]/card[P^*(X)]$. Fig. 2 shows the relationship of the lower and upper approximation of an information system.

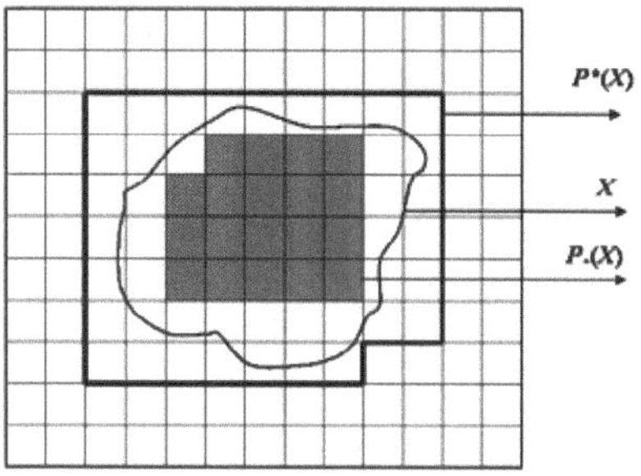

P*(X)- Upper approximation, P*(X)- Lower approximation, X – Boundary region
Fig.1: Rough approximation set

REDUCTION OF ATTRIBUTES AND CORE

A ***reduct*** is a subset of attributes which by itself can fully characterize the knowledge in the database [16]. The subset $R \subseteq P \subseteq Q$ such that $\gamma_P(Y) = \gamma_R(Y)$ is called Y-reduct of P and denoted as $Red_Y(P)$. The *core*is possessed by every legitimate *reduct* and cannot be removed from the information system without deteriorating basic knowledge of the system. The *Y-core* is intersection of all *Y-reducts* of P, included in every *Y-reducts* of P as: $Core_Y(P) = \cap Red_Y(P)$.

SIGNIFICANCE OF ATTRIBUTES

A set of attribute D is depends totally on a set of attributes C, denoted as $C \Rightarrow D$, if all values of attributes from D are uniquely determined by values of attributes from C. Let D and C be a subset of A, the significance factor of an individual attribute a $\in$ C with respect to dependency between D and C is represented as: $\sigma(a, C, D) = \gamma_C(U/D) - \gamma_{P-\{a\}}(U/D)$.

DISCUSSION

The results of jackknife tests, it is observed that α/β class has the highest accuracy irrespective of whichever class or algorithms. This may be related to the proportion of α/β class in the training sets in which α/β class occupied the biggest part, as shown in Fig. 2. As a supervised learning method, it makes it easier to capture characteristics that feed more training objects to Rough Sets.

Although the average accuracy of 498 domains of Rough Sets is slightly lower than the Support Vector Machine, they are still much better than others. So, from the results of jackknife, the conclusion is that the performance of Rough Sets should have exceeded the component-coupled algorithm and neural networks in this study, and parallel with SVM algorithm. In addition, since the extraction of data, coupling effects among amino acids are not considered yet, only the amino acid composition and 8 physicochemical properties are taken into the account, which may influence the secondary structure of proteins. Based on this point, it is reasonable to believe that the algorithm based on Rough Sets still has potential to improve. However, a quick search of biological literatures shows that Rough Sets are still seldom used in bioinformatics, except for some applications in medical and health related fields. One obstacle for the application of Rough Sets in bioinformatics may be the large amounts of biological data and the comparatively slow computational speed of Rough Sets algorithm.

There are several factors that may affect the precision of prediction based on Rough Sets. One of them is the conditional attributes, and another is the scale of datasets. The selection of conditional attributes must reflect the relation between sequences and their structural classes. If a set of conditional attributes does not make this bridge, Rough Sets can not induce effective rules from the decision system. From this study, the amino acid compositions and physicochemical properties certainly can be used to discriminate protein sequences from different structural classes. However, the conformation of secondary structure of protein is very complex, and there are still other factors that may influence this process

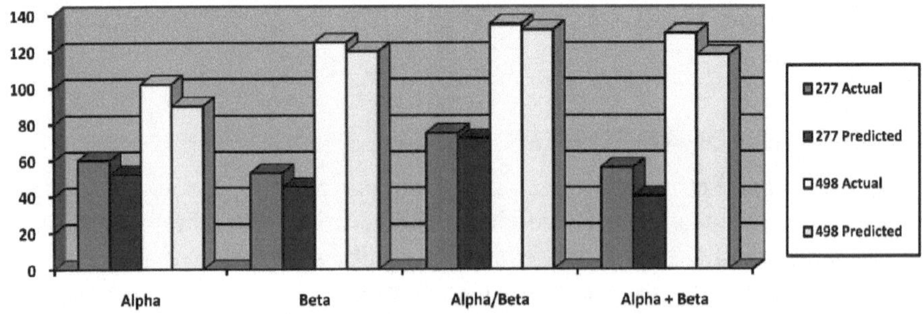

Fig. 2: The comparisons between actual objects number and predicted number of 4 protein structural classes.

and that can be taken into account to improve this method. In theory, the more objects in the dataset, the more accurate the prediction would be; in other words, the more information there is about the problem, the more likely to induce useful rules from it. It is observed that the accuracy of 498 domains is much higher than the 277 domains.

METHODS

Two datasets of protein domain sequences used in this study came from Zhou [16], which was extracted respectively from SCOP database [27], one consisted of 277 sequences, and another 498 sequences. The composition of 4 domain classes is listed in the Table 1 respectively. Two datasets are used to test this proposed method through self-consistency test, jackknife test, and induced decision rules from datasets, and compared the total accuracy and accuracies of each structural class with other algorithms.

CONSTRUCTION OF DECISION SYSTEM

The 3D structure of proteins is uniquely determined by their amino acid sequences, i.e. the primary sequences. In order to predict the structural classes of proteins, primary sequences have been converted into numerical vectors, in other words, features information has been extracted from primary sequences as the representation of proteins. There are many ways to extract features from amino acid sequences for representing the essential characteristics of proteins. In this paper, the amino acid compositions and physico-chemical properties are used as the features to describe amino acid sequences, i.e. conditional attributes set. Because of the degenerate nature of the sequence-structure relationship, the number of their folding patterns is quite limited, despite that the number of protein sequences is extremely large. In fact, according to their chain folding patterns, proteins are usually folded into one of the following four structural classes: all-α, all-β, α/β, and $\alpha+\beta$.

Given a decision system $A = (U, A \cup \{d\})$, here A is called the conditional attributes set, and $d \notin A$ is called decision attribute. The elements of the universe U are called objects. Each object in U is in accordance with a decision value with it in $\{d\}$. In this work, conditional attributes set A is made up of two parts, the first part is S_1 = {A, R, N, D, C, Q, E, G, H, I, L, K, M, F, P, S, T, W, Y, V}, which is composed of compositional percentages of the 20 amino acids in primary sequences, and the second part is S_2 = {KR, ED, AGP, KRED, KR-ED, FIKMNY, LVIFM, ST}, the 8 physicochemical properties: positive charge (KR), negative charge (ED), total charge (KRED), net charge (KR-ED), major hydrophobic (LVIFM), and groupings ST, AGP, and FIKMNY. It can be written as: $A = S_1 \cup S_2$, and the decision attribute {d} = {all-α, all-β, α/β, $\alpha+\beta$}.

MODEL INDUCTION

Rough sets based models are built on the concept of indiscernibility relation. Given a decision system A= (U, A∪{d}), it is defined that $IND_A(A,x,d)$ to be the set of objects that are indiscernible from x with respect to the decision attribute d. From the definition of indiscernibility we derive for each object x ∈ U the set of reducts $RED_A(x,d)$ to be the set of minimal sets of attributes B⊆A such that $IND_A(B,x,d) = IND_A(A,x,d)$. Hence, a reduct of x is a minimal set of attributes B with the same discriminatory power as A. Finding the set of all reducts is a NP-hard problem, however, there are heuristics that compute a sufficient number of reducts in an acceptable computing time. Since real-world data almost always is polluted with noise, methods finding approximate reducts that reveal the underlying, general pattern in the data have also been developed. And the NP-hard problem referred can itself be considered with further time-complexity growth-rate function theories.

MODEL EVALUATION AND APPLICATION

In a self-consistency test, training sets will be predicted with those decision rules trained by them. The accuracy of self-consistency test can tell us how well the rules captured the characteristics of training sets. As elucidated in a comprehensive review [28], jackknife test is deemed to be the most objective and rigorous way to estimate the performance of a classifier. Each object in the dataset will in turn be the test set, and the left as training set. In other words, each protein of the dataset will be predicted once by the classifier trained with the left proteins. After this process, we can average all iterations to obtain an unbiased number of performance estimates. In this study, self-consistency tests and jackknife tests are performed against two datasets, and the results are compared with other algorithms, including component coupled algorithm [16], neural networks [29] and SVM [11].

ALGORITHM AND IMPLEMENTATION

Except for several Perl scripts we wrote to deal with protein sequences and extract features from them, all of computation concerning on Rough Sets is implemented on Rosetta system, which is a toolkit for data mining and knowledge discovery using rough sets. Rosetta system is implemented with C++, and is ported to a variety of platforms. More detailed descriptions and cases of application on Rosetta system can be found in Øhrn's Ph.D dissertation [17]. Rosetta system includes a large variety of typical algorithms that can be used in almost each step of data mining with rough sets. Users can choose appropriate algorithms for various situations. Another advantage of Rosetta system is that users can perform very complicated cross-validation tests based on a user defined script. In this study, a pipeline script is written for performing *CV Serial Executor* test on Rosetta (Fig. 3) for cross-validation. This *CV Serial Executor* is an option available in Rosetta software to execute user's own scripts. With this script, the jackknife test of cross-validation over the datasets as described is performed. For our datasets, two critical steps that may affect predicting results are discretization and reduction. In this work, the Semi-Naive algorithm (SemiNaiveScaler) and genetic algorithm (SAVGeneticReducer) respectively for the two steps are used.

```
%Training pipeline.
SemiNaiveScaler
{MODE=Save; MASK = True; FILENAME = d:\temp\cuts#ITERATION#.txt}
EqualFrequencyScaler
{MODE = Save; MASK = True; FILENAME = d:\temp\backupcuts#ITERATION#.txt}
SAVGeneticReducer
{DISCERNIBILITY = Object; MODULO.DECISION = True; SELECTION = All;
cost = F;
APPROXIMATE = True; FRACTION = 0.95; KEEP.LEVELS = 1}
RuleGenerator
{}
%Testing pipeline
OrthogonalFileScaler
{MODE = Load; MASK = False; FILENAME = d:\temp\cuts#ITERATION#.txt}
OrthogonalFileScaler
{MODE = Load; MASK = False; FILENAME = d:\temp\backupcuts#ITERATION#.txt}
BatchClassifier
{CLASSIFIER = StandardVoter; FALLBACK = True; FALLBACK.CLASS = Yes;
FALLBACK.CERTAINTY = 0.3; MULTIPLE = Best; LOG = True;
LOG.FILENAME = d:\temp\log#ITERATION#.txt; LOG.VERBOSE = True;
CALIBRATION = False; ROC = False;}
```

Fig.3: Pipeline script for cross-validation test on Rosetta system

As discussed above, computing all reducts of a decision table is a NP-hard problem. Many alternative algorithms have been developed for computing of approximate reducts.

CONCLUSION

In summary, a method based on Rough Sets theory is reported to predict protein structural classes. Two datasets constructed by Zhou [17] are used to test this method. High prediction accuracies have been achieved through self-consistency and cross-validation test. This suggests that the rough sets approach holds a high potential to become a useful tool in bioinformatics. Furthermore, as elucidated in a recent review by Chou [19], the past progresses in protein structural class prediction had a series of impacts to the development for predicting many other attributes of proteins. It is believed that the current rough sets approach might also stimulate the development in predicting other protein attributes, such as subcellular location [20-22], membrane protein type [23-25], and enzyme family classification [26], among many others.

REFERENCES

1. Cai, Y.D., Liu, X.J., Xu, X. and Zhou, G.P. 2001. Support vector machines for predicting protein structural class. *BMC Bioinformatics*, 2(1):3.

2.Cai, Y and Zhou, G: 2000. Prediction of protein structural classes by neural network. *Biochimie*, 82(8):783-785.

3.Chou, K.C. 2005. Progress in protein structural class prediction and its impact to bioinformatics and proteomics. *Curr Protein Pept Sci*, 6(5):423-436.

4.Chou, K.C., Cai and Y.D. 2004. Predicting protein structural class by functional domain composition. *Biochem Biophys Res Commun*, 321(4):1007-1009.

5.Chou, K.C. and Cai, Y.D. 2004. Predicting enzyme family class in a hybridization space. *Protein Sci*, 13(11):2857-2863.

6.Chou, K.C. 2001. Prediction of protein cellular attributes using pseudo-amino acid composition. *Proteins*, 43(3):246-255.

7.Chou, K.C. 2000. Prediction of protein structural classes and subcellular locations. *Curr Protein Pept Sci*, 1(2):171-208.

8.Chou, K.C. and Elrod, D.W. 1999. Protein subcellular location prediction. *Protein Eng*, 12(2):107-118.

9.Zhou, G.P. and Doctor, K. 2003. Subcellular location prediction of apoptosis proteins. *Proteins*, 50(1):44-48.

10.Chou, K.C. and Elrod, D.W. 1999. Prediction of membrane protein types and subcellular locations. *Proteins*, 34(1):137-153.

11.Chou, K.C. and Maggiora, G.M. 1998. Domain structural class prediction. *Protein Eng*, 11(7):523-538.

12.Chou, K.C. 1995. A novel approach to predicting protein structural classes in a (20-1)-D amino acid composition space. *Proteins,* 21(4):319-344.

13.Chou, K.C. and Zhang, C.T. 1995. Prediction of protein structural classes. *Crit Rev Biochem Mol Biol*, 30(4):275-349.

14.Chou KC and Zhang, C.T. 1994. Predicting protein folding types by distance functions that make allowances for amino acid interactions. *J Biol Chem*, 269(35):22014-22020.

15. Feng, K.Y., Cai, Y.D. and Choum K.C. 2005. Boosting classifier for predicting protein domain structural class. *Biochem Biophys Res Commun*, 334(1):213-217.

16. Klein, P: 1986. Prediction of protein structural class by discriminant analysis. *Biochim Biophys Acta*, 874(2):205-215

17. Klein, P. and Delisi, C. 1986. Prediction of protein structural class from the amino acid sequence. *Biopolymers,*25(9):1659-1672.

18. Metfessel, B.A., Saurugger, P.N., Connelly, D.P. and Rich, S.S.: 1993. Cross-validation of protein structural class prediction using statistical clustering and neural networks. *Protein Sci*, 2(7):1171-1182.

19. Murzin, A.G., Brenner, S.E., Hubbard, T. and Chothia, C. 1995. SCOP: a structural classification of proteins database for the investigation of sequences and structures. *J Mol. Biol.*, 247(4):536-540.
20. Ohrn, A. Discernibility and Rough Sets in Medicine: 1999. Tools and Applications. In Department of Computer and Information Science Volume PhD. Norway , *Norwegian University of Science and Technology*: 239.
21. Pawlak Z, Rough sets: 1991. theoretical aspects of reasoning about data. In Theory and decision library Series D, System theory, knowledge engineering, and problem solving; v 9 Dordrecht ; Boston , Kluwer Academic Publishers: xvi, 229.
22. Pawlak, Z., Rough sets: 1982. *International Journal of Computer and Information Sciences*, 11:341-356.
23. Shen, H.B., Yang, J., Liu, X.J. and Chou, K.C. 2005 Using supervised fuzzy clustering to predict protein structural classes. *Biochemical and Biophysical Research Communications*, 334(2):577.
24. Shen, H.B., Yang, J. and Chou, K.C. 2005. Fuzzy KNN for predicting membrane protein types from pseudo-amino acid composition. *J. Theor Biol.*
25. Wang, M., Yang, J., Xu, Z.J. and Chou, K.C. 2005. SLLE for predicting membrane protein types. *J Theor Biol*, 232(1):7-15.
26. Zhang, C.T. and Chou, K.C. 1992. An optimization approach to predicting protein structural class from amino acid composition. *Protein Sci*, 1(3):401-408.
27. Zhou, G.P. and Assa-Munt N: 2001. Some insights into protein structural class prediction. *Proteins,* 44(1):57-59.
28. Zhou, G.P. 1998. An intriguing controversy over protein structural class prediction. *J. Protein Chem*, 17(8):729-738.
29. Zhou, G., Xu, X. and Zhang, C.T. 1992. A weighting method for predicting protein structural class from amino acid composition. *Eur. J. Biochem*, 210(3):747-749.

27

Plant Systems Biology

G. Ramesh Kumar, CP Rajadurai and M. Ramachandran

SYSTEMS BIOLOGY

The ultimate goal of any biological experiment is to understand biological systems in sufficient detail to enable accurate, quantitative predictions about the behavior of the systems, including predictions of the effects of modifications of the systems. To precisely simulate the behavior of the system using a computer model is one of the ways to achieve the above goal and in this venture we are nowhere close to the goal. Systems biology will help us to get closer to this goal and it is expected that biologists to adopt to take advantage of the power of systems biology approach. Systems biology is an emergent field that aims at system-level understanding of biological systems. Systems Biology is the study of the interactions between the components of biological systems, and how these interactions give rise to the function and behavior of that system. For example, theenzymes and metabolites in a metabolic pathway.With the progress of genome sequence projects and range of other molecular biology projects that accumulate in-depth knowledge of molecular nature of biological system, we are now at the stage to seriously look into possibility of system-level understanding of solidly grounded on molecular-level understanding.Unlike molecular biology which focus on molecules, such as sequence of nucleic acids and proteins, systems biology focus on systems that are composed of molecular components. Although systems are composed of matters, the essence of system lies in dynamics and it cannot be described merely by enumerating components of the system.

OBJECTIVES

1. To understand the structure of the system, such as gene regulatory and biochemical networks, as well as physical structures.
2. To understand the dynamics of the system, both quantitative and qualitative analysis as well as construction of theory/model with powerful prediction capability.
3. To understand the control methods of the system.
4. To understand the design methods of the system, are key milestones to judge how much we understand the system.

DEFINITIONS

Many definitions have been given to study Systems Biology. They are as follows:

- Integrative approach in which scientists study pathways and networks will touch -all areas of biology, including drug discovery (American Chemical Society).
- Systematic approach not focused on individual genes and individual proteins, instead interested in analyzing whole systems of genes or proteins by capturing information from many different elements of the overall system (The Stuttgart University).
- Leroy Hood's definition is that it involves 'integrating technology, biology and computation'.
- A hopefully uncontroversial definition would be "Integrative approaches in which scientists study and model pathways and networks, with interplay between experiment and theory".
- According to Lauffenburger systems biology can be referred to the "four M's:"
    - Measurement,
    - Mining,
    - Modelling, and
    - Manipulation.

Manipulation and measurement are on the experimental side. Mining and modelling are on the computational side. These four M's are part of an iterative process, beginning with manipulating the system. Once a system is perturbed, it is measured using a high-throughput, multivariate technology. The data are then mined to elucidate hypotheses that, when cast in terms of formal computational models, form the basis for a new manipulation of the system. Because modelling is an integral part of systems biology, many computational tools are required.

RELATED DISCIPLINES

According to the interpretation of System Biology as the ability to obtain, integrate and analyze complex data from multiple experimental sources using interdisciplinary tools, some typical technology platforms required are:

- **Phenomics**: Organismal variation in phenotype as it changes during its life span.
- **Genomics:** Organismal deoxyribonucleic acid (DNA) sequence, including intra-organisamal cell specific variation.
- **Epigenomics / Epigenetics**: Organismal and corresponding cell specific transcriptomic regulating factors not empirically coded in the genomic sequence. (i.e. DNA methylation, Histone Acetelation etc.).
- **Transcriptomics:** Organismal, tissue or whole cell gene expression measurements by DNA microarrays or serial analysis of gene expression(SAGE).
- **Interferomics:** Organismal, tissue, or cell level transcript correcting factors (i.e. RNA interference).
- **Translatomics / Proteomics:** Organismal, tissue, or cell level measurements of proteins and peptides via two-dimensional gel electrophoresis, mass spectrometry or multi-dimensional protein identification techniques.
- **Metabolomics:** Organismal, tissue, or cell level measurements of all small-molecules known as metabolites.
- **Glycomics:** Organismal, tissue, or cell level measurements of carbohydrates.
- **Lipidomics:** Organismal, tissue, or cell level measurements of lipids. In addition to the identification and quantification of the above given molecules further techniques analyze the dynamics and interactions within a cell.
- **Interactomics:** Organismal, tissue, or cell level study of interactions between molecules. Currently the authoritative molecular discipline in this field of study is protein-protein interactions.
- **Fluxomics:** Organismal, tissue, or cell level measurements of molecular dynamic changes over time.
- **Biomics:** Systems analysis of the biome.

The investigations are frequently combined with large scale perturbation methods, including gene-based and chemical approaches using small molecule libraries. Robots and automated sensors enable such large-scale experimentation and data acquisition. These technologies are still emerging and many face problems that the larger the quantity of data produced, the lower the quality. A wide variety of quantitative scientists are working to improve the quality of these approaches and to create, refine, and retest the models to accurately reflect observations.The systems biology approach often involves the development of mechanistic models, such as the reconstruction of dynamic systems from the quantitative properties of their elementary building block. Due to the large number of parameters, variables and constraints in cellular networks, numerical and computational techniques are often used. Other aspects of computer science and informatics are also used in systems biology.The term systems biology has widespread currency and popularity as of 2007 with chairs and institutes of systems biology proliferating worldwide (Table 1).

Table 1: List of systems biology institute and research centre originated worldwide

Institute/Centre	Web Link
Institute of Bioinformatics and Systems Biology	http://bioinfo.nctu.edu.tw/departmentinfo/pages. php? ID= about
Systems Biology at the Pacific Northwest National Laboratory	http://www.sysbio.org/
The Swiss Initiative in Systems Biology	http://www.systemsx.ch/
Semantic Systems Biology	http://www.semantic-systems-biology.org/
Systems Biology Portal	http://www.systems-biology.org/
Systems Biology Biochemweb	http://www.biochemweb.org/systems.shtml
Institute of Systems Biology	http://www.systemsbiology.org/
Plant Systems Biology	http://www.psb.ugent.be/
Plant Systems Biology at UCSD	http://www.biology.ucsd.edu/psbigert/

RENAISSANCE OF PLANT RESEARCH

AGRO-NOMICS

ARABIDOPSIS GROWTH NETWORK INTEGRATING OMICS TECHNOLOGIES

Plants are crucial to mankind as they supply food, chemicals, pharmaceuticals and renewable sources of materials and energy. Yet the processes by which they grow are poorly understood. Although some of the key factors involved in plant organ growth have already been identified, the circuitry that links the different levels of organization (whole plant, organ, cell, molecular module, and molecule) remains to be uncovered. Fortunately, for the first time, techniques exist to characterize multicellular system exhaustively at all relevant levels. The main goals of Agro-Nomics are:

1. To survey systematically with an array of high throughput methods what are the molecular components driving growth in the cells of a developing *Arabidopsis* leaf,

2. To understand how these elements interact and coordinate their action across levels of organisation, and
3. To explain quantitative growth phenotypes at the molecular level through inference and mathematical modelling, followed by further experimentation.

PLANT SYSTEMS BIOLOGY

Plants have played a major role in the geochemical and climatic evolution of our planet. Today, in addition to their fundamental ecological importance plants are essential for humans as the main source of food, provide raw materials for many types of industry and chemicals for medical applications. It is thus daunting to realize how little we understand about plant systems. To date, only approximately *Arabidopsis thaliana*, the most explored model system for plant biologists, have been characterized experimentally. Systems biology offers the opportunity to increase our understanding of plants as living organisms, by generating a holistic view of the organism grounded at the molecular level. In this chapter, we discuss the basics of systems biology, the data and tools we need for systems research and how it can be used to produce an integrated view of plant biology.

COMPLEXITY AND ROBUSTNESS IN BIOLOGICAL SYSTEMS

Everyone agrees that biological systems are complex. But what does complexity mean? Is complexity in biological systems just a consequence of the inherent difficulty in understanding them? We refer to a system as a complex when it exhibits the following characteristics:

1. It is composed of many different elements.
2. The constituent elements interact.
3. The elements and their interactions are dynamic, and are often governed by non-deterministic rules.
4. The elements interactions can be influenced by external factors.

As a consequence of these properties, complex systems exhibit unexpected or emergent behaviour that cannot be understood from studying the parts in isolation. In addition, mathematical models that describe them usually involve nonlinear behaviour. Even the simplest life forms such as *Mycoplasma* are complex by this definition. Living systems are composed of a large number of elements with diverse chemical properties that are intricately interconnected. These elements and connections are dynamic and respond to internal or external factors. For instance, the messenger RNA levels of hundreds of genes that code for hundreds of different proteins can vary over time due to the action of signal transduction cascades activated by developmental or environmental cues. In addition to complexity, and perhaps because of it, biological systems are robust. This key and ubiquitous feature of biological systems refers to their ability to maintain proper functions despite internal and external perturbations and uncertainty . Robustness is an example of a property that emerges at the system level and that cannot be understood at the individual part level.

The types of perturbations encountered by living systems are varied and include: stochastic noise, physiological and developmental signals, environmental change and genetic variation. It is important to distinguish robustness from homeostasis.In contrast, homeostasis refers to maintaining the state of the system. A system is robust as long as it maintains functionality, even if this is achieved by moving to a new steady state and for instance, during the diauxic shift yeasts drastically changes from anaerobic to aerobic metabolism. This transition involves genome-wide changes in gene expression and the reprogramming of metabolic pathways. In this example, the state of the system is dramatically altered and however, the system is robust as it continues to produce adenosine triphosphate and grow. Robustness allows for changes in the structure and components of the system in response to perturbations, but maintains specific functions.

GENERAL PRINCIPLES

When networks, or graphs in a more formal mathematical language, are applied to molecular biology nodes (or vertices) represent the molecules present inside a cell (e.g. proteins, RNAs and/or metabolites) and links (or edges) between nodes represent their biological relationships (e.g. physical interaction, regulatory connections and metabolic reactions). For example, in protein interaction networks, protein complexes can be represented as networks where nodes represent proteins and their edges represent the physical interactions between them. In metabolic networks, nodes can represent metabolites and the edges the metabolic reactions that transform one metabolite into another. In genetic networks, nodes can represent genes and the link between them a genetic interaction, such as synthetic lethality. But other abstractions are possible and for instance, nodes can also be used to represent biological processes and the edges connecting them indicate the functional relationship between the processes. This network simplification has uncovered basic principles of the structure and organization of different types of molecular networks in many different organisms and has been successful for biological research.

EXPERIMENTAL APPROACHES FOR PLANT SYSTEMS BIOLOGY

A systems biology approach is possible today because of the breakthrough in technologies that allow us to measure and connect the entire complement of defined molecules inside an organism. The new influx of high-throughput data is shifting biology from a reductionist to a systems-level view. Practically speaking, systems biology generates and integrates different types of genome-wide data in living organisms to produce models of the system as a whole. Ideker *et al.*, described a fundamental framework to carry out research in systems biology:

1. Define all the components of the system,
2. Systematically perturb and monitor components of the system,
3. .Reconcile the experimentally observed responses with those predicted by the model and
4. Design and perform new perturbation experiments to distinguish between multiple or competing hypotheses.

THE GENOME

The first step to understand any kind of plant from a systems point of view is to enumerate its basic components. The genome of an organism is the complete hereditary information encoded in the DNA (or RNA in some viruses). The first genome of a complex free-living organism to be completed was that of *Haemophilus influenza*. The tremendous advances in sequencing technologies since then, has catapulted us into the genomic era. Today, the scientific community benefits from the complete genome sequences of many organisms including several plants. In 2000, the complete genome sequence of the plant *A. thaliana* was published and in 2005, the first monocotyledoneous plant *Oryza sativa* was sequenced. In addition, at the beginning of 2008, the first genome sequence of a genetically modified plant was reported, for the SunUp *Carica papaya*. Genomes provide the parts lists as far as genes and their products (proteins and RNAs) are concerned.

There are two main approaches for 'sampling' the gene space of genomes without sequencing them completely:

1. Sequencing expressed sequence tags (ESTs) and
2. Sequencing genomic sequences that have been filtered based on either hypomethylation or High-Cot.

EST sequencing involves preparing cDNA of the transcribed messages and then randomly sequencing them from either the 3 prime or the 5 prime. The advantage of EST sequencing is that the sequences obtained represent the transcribed regions of the genome. The disadvantages are that ESTs typically do not represent the entire cDNA (only about 500 bp) and they tend to represent only roughly 50% of the total genes in the genome. Large stretches of repetitive DNA and transposable elements tend to be hypermethylated compared to gene coding regions which are unmethylated, for example 95% of the maize exons are thought to be unmethylated. Genomic libraries can be prepared enriched for unmethylated DNA by cloning in bacterial hosts that destroy methylated DNA. Alternatively, the High-Cot method removes repetitive DNA by depending on the rapid association of repetitive DNA. The advantage of sequencing filtered genomic sequences over EST sequencing is that it is possible to obtain the introns and regulatory regions of the gene rather than just the transcribed portion of the gene.

Today, many specialized software are available in the market as well as freeware. Even though in silico predictions are an excellent start, they have to be used critically because they are not always accurate. Indeed, gene predictions and genome annotations are typically dynamic improving over time. For example, at least 40% of the original predictions made for the *Arabidopsis* genome were subsequently found to be wrong and corrected in later releases of the annotation. The next step after gene discovery is gene functional annotation and this functional analysis can be done in several ways: computational predictions, gene expression, mutant analysis and protein–protein interactions, just to name some of the most common approaches. However, high-throughput screens of mutants are fundamental for accessing plant gene functions on a genomic scale.

SYSTEMS BIOLOGY IN *ARABIDOPSIS* RESEARCH

The development of systems biology approaches and tools for *Arabidopsis* has enabled the generation of testable biological hypotheses derived from the integration and analysis of genomic data within a network context and it has uncovered testable hypotheses for regulatory networks in Arabidopsis from the following two studies,

1. Qualitative network models and genome-wide expression data define carbon/nitrogen molecular machines in *Arabidopsis.*
2. A systems approach identifies an organic nitrogen-responsive gene network regulated by the master clock gene CCA1.

ADVANCED COMPUTATIONAL TECHNIQUES

There are several different computational techniques that have been developed for processing of data, such as images and experimental measurements, and construction of simulation models and one such model is reaction-diffuse where the pattern forming substance is assumed to diffuse across cells. Meyerowitz and colleagues used this model to study the shape and size of the shoot apical meristem (SAM) and expression domains of the WUSCHEL (WUS) gene.WUS is known to be involved in SAM development and is known to induce the ligand CLAVATA3 (CLV3) which binds to the CLAVATA1 (CLV1) receptor kinase. CLV1 then activates a signalling cascade that represses WUS. In this study, the authors used in vivo confocal microscopy and image-processing algorithms to obtain quantitative measurements of gene expression. The measurements were subsequently used as input for the reaction-diffuse model, which allowed them to predict the outcome of laser cell ablation experiments and removal of CLV3 signal. The model correctly simulated creation of two new WUS domains after the ablation of the central cells from the SAM and the expansion of WUS expression with the removal of CLV3.

COMPLETED LARGE-SCALE PLANT GENOME SEQUENCING PROJECTS

Arabidopsis thaliana 5 chromosomes,plastid, mitochondrion

Glycine max 20 chromosomes,plastid, mitochondrion

Medicago truncatula,8 chromosomes,plastid, mitochondrion

Oryza sativa 12 chromosomes,plastid, mitochondrion, mitochondrial plasmid B1, mitochondrial plasmid B2

Populus trichocarpa 19 chromosomes,plastid, mitochondrion

Sorghum bicolor 10 chromosomes,plastid, mitochondrion

Vitis vinifera 19 chromosomes,plastid, mitochondrion

Zea mays 10 chromosomes,plastid, mitochondrion, mitochondrial 1.9 kb plasmid

IN-PROGRESS LARGE-SCALE PLANT GENOME SEQUENCING PROJECTS

Brachypodium distachyon 5 chromosomes, plastid, mitochondrion

Carica papaya 9 chromosomes, plastid, mitochondrion

Lotus japonicus 6 chromosomes, plastid, mitochondrion

Manihot esculenta 20 chromosomes, plastid, mitochondrion

Solanum lycopersicum 12 chromosomes, plastid, mitochondrion

Solanum tuberosum 12 chromosomes, plastid, mitochondrion

SOFTWARE TOOLS FOR PLANT SYSTEMS BIOLOGY

SYSTEMS BIOLOGY WORKBENCH

The Systems Biology Workbench (SBW) is a modular, broker-based, message-passing framework for communication between applications that aid in research in systems biology.

FEATURES INCLUDE

1. Bindings for C, C++, Delphi, Java, Perl, Python
2. Support for Windows, FreeBSD and Linux (with Mac OS X planned)
3. Distributed operation via SSH
4. Bidirectional CORBA-SBW gateway
5. Collection of modules provided with the base distribution:

6. Simple stochastic simulator
7. An SBML-to-MATLAB ODE and Simulink file translator
8. An SBML reader module
9. A "clipboard" module for exchanging SBML models
10. A Browser that outputs module interface definitions
11. A simple plotting module for time-series data
12. A generic simulation-control GUI interface

SBML (The Systems Biology Markup Language) is a computer text language that facilitates the transfer of computational models. People from around the world participated in the development of SBML. The project was conceived by Hamid Bolouri, now at the Institute for Systems Biology, and Hiroaki Kitano, head of the ERATO Kitano Symbiotic Systems Project. Another part of the project is the Systems Biology Workbench, which is intended to foster the reuse of software and to allow people to build upon what has already been created.

The Systems Biology Markup Language (SBML) is a computer-readable format for representing models of biochemical reaction networks.

SBML is applicable to:

1. Metabolic networks
2. Cell-signaling pathways
3. Genomic regulatory networks
4. Many other areas in systems biology
5. SBML is supported by over 100 different software applications used in systems biology research.

For more information about SBML, see http://sbml.org.

The large-scale genomic data available to biologists today require the use of a variety of software tools to process and analyze the data. A systems approach to understanding biology involves an iterative process of data integration, building a model, designing experiments to support the model and generating new hypothesis based on new data. There are several tools that are specific for each stage, however, few of them span across several stages and even fewer allow for the iterative analysis. Several softwares and tools have been developed in the past years that permit data integration and biological system modeling (Table 2).

Table 2: List of systems biology software for studying biological networks

Software / Tool	*Developer*
Sungear	Poultney et al., 2007
Genevestigator	Zimmermann et al., 2004
MapMan	Thimm et al., 2004
Cytoscape	Shannon et al., 2003
Gepasi	Mendes, 1997
Virtual Cell	Loew and Schaff, 2001
Osprey	Breitkreutz et al., 2003
Arabidopsis eFP browser	Winter et al., 2007

Such software allows detailed mathematical representation of cellular processes as well as qualitative representations of cellular components and their interactions.Due to the increased use of networks to represent interaction data, viewers and graph drawing software such as Pajek, daVinci,Graphlet and Graphviz developed for general purpose graph drawing and visualization, are now used to organize and display biological data as graphs. Other advancement of such a software include various methods to layout the network graphs, support connection to external databases and include more sophisticated data analysis tools.

Web sources for systems biology

1. CellDesigner — ERATO Kitano Symbiotic Systems Project http://www.systems-biology.org/002/
2. Cellerator — NASA JPL & University of California Irvine
3. Cytoscape — Institute for Systems Biology & MIT
4. Gepasi — Virginia Tech http://gepasi.dbs.aber.ac.uk/softw/gepasi.html
5. Jarnac — Keck Graduate Institute
6. JDesigner — Keck Graduate Institute
7. JigCell — Virginia Tech
8. NetBuilder — University of Hertfordshire
9. SigPath — Mount Sinai
10. StochSim — Cambridge University
11. Virtual Cell — University of Connecticut Health Center
12. WinSCAMP — Keck Graduate Institute
13. On going :E-CELL, Dbsolve, GSK PWF, more

Modelling

1. Cellerator
2. E-cell
3. Gepasi
4. Jarnac
5. Stochsim
6. Triple-J
7. Yeast Model

Markup languages and standards

1. Biopax
2. CellML
3. CML
4. PEDRo
5. SBML

Other useful tools

1. Biolayout - Bio Sketch Pad - Gene Mapp

REFERENCES

Adams, M.D., Kelley, J.M., Gocayne, J.D., Dubnick, M., Polymeropoulos, M.H., Xiao, H., et al. 1991. Complementary, D.N.A sequencing: expressed sequence tags andhuman genome project. *Science* 252:1651-1656.

Alonso, J.M. and Ecker, J.R. 2006. Moving forward in reverse: genetic technologiesto enable genome-wide phenomic screens in Arabidopsis. *Nat Rev Genet.*,7:524-536.

Barabasi, A.L. and Oltvai, Z.N. 2004. Network biology: understanding the cell'sfunctional organization. *Nat Rev Genet* 5:101–113.

Batagelj, V. and Mrvar, A. 1998. Pajek – program for large network analysis. *Connections*, 21:47-57.

Bennet, M.D. and Leitch, I.J. 1995. Nuclear DNA amounts in angiosperms. *Ann Bot.*, 76:113-176.

Bennetzen, J.L., Schrick, K., Springer, P.S., Brown, W.E. and SanMiguel, P. (1994).Active maize genes are unmodified and flanked by diverse classes of modified,highly repetitive DNA. *Genome,* 37:565–576.

Breitkreutz, B.J., Stark, C. and Tyers, M. 2003. Osprey: a network visualization system. *Genome Biol* 4:r22.1–r22.4.

DeRisi, J.L., Iyer, V.R. and Brown, P.O. 1997. Exploring the metabolic and geneticcontrol of gene expression on a genomic scale. *Science* 278:680-686.

di Bernardo, D; Thompson, M.J., Gardner, T.S., Chobot, S.E., Eastwood, E.L., Wojtovich, A.P., Elliot, S.J., Schaus, S.E. and Collins, J.J. 2005. Chemogenomic profiling on a genome-wide scale using reverse-engineered gene networks. *Nature Biotechnology* 23:377-383.

Fletcher, J.C., Brand, U., Running, M.P., Simon, R. and Meyerowitz, E.M. 1999. Signaling of cell fate decisions by CLAVATA3 in Arabidopsis shoot meristems. *Science*, 283:1911-1914.

Fleischmann, R.D., Adams, M.D., White, O., Clayton, R.A., Kirkness, E.F., Kerlavage, A.R., et al. 1995. Whole-genome random sequencing and assembly of Haemophilusinfluenzae Rd. *Science*, 269:496-512.

Gardner, T.S., di Bernardo, D., Lorenz, D. and Collins JJ 2003. Inferring genetic networks and identifying compound of action via expression profiling. *Science*, 301:102-1005.

Gutierrez, R.A., Shasha, D.E. and Coruzzi, G.M. 2005. Systems biology for the virtualplant. *Plant Physiol.*, 138:550-554.

Gierer, A. and Meinhardt, H. 1972. A theory of biological pattern formation. *Kybernetik*, 12:30-39.

Gloria, M. Coruzzi, Alejandro, R. Burga, Manpreet, S. Katari and Rodrigo A. Guti´errez 2008. Systems Biology: Principles and Applications in Plant Research. *Annu.Plant Rev.*, 16(4):1-40.

Jonsson, H., Heisler, M., Reddy, G.V., Agrawal, V., Gor, V., Shapiro, B.E., et al. (2005)Modeling the organization of the WUSCHEL expression domain in the shoot apicalmeristem. *Bioinformatics* 21 (Suppl 1): i232–i240.

Kitano, H. 2002. Computational systems biology. *Nature* 420:206–210.

Kitano, H. 2004. Biological robustness. *Nat Rev Genet.*, 5:826–837.

Kitano, H. 2007. Towards a theory of biological robustness. *Mol. Syst. Biol.* 3:137.

Locke, J.C., Southern, M.M., Kozma-Bognar, L., Hibberd, V., Brown, P.E., Turner, M.S.,et al. 2005. Extension of a genetic network model by iterative experimentation andmathematical analysis. *Mol. Syst. Biol.* 1:0013.

Loew, L.M. and Schaff, J.C. 2001. The virtual cell: a software environment for computational.

Mendes, P. 1997. Biochemistry by numbers: simulation of biochemical pathways withGepasi 3. *Trends Biochem Sci.*, 22:361-363.

Merchant, S.S., Prochnik, S.E., Vallon, O., Harris, E.H., Karpowicz, S.J.,Witman, G.B.,et al. 2007. The Chlamydomonas genome reveals the evolution of key animal andplant functions. *Science*, 318:245-250.

Ming, R., Hou, S., Feng, Y., Yu, Q., Dionne-Laporte, A., Saw, J.H., et al. 2008. Thedraft genome of the transgenic tropical fruit tree papaya (Carica papaya Linnaeus). *Nature*, 452:991-996.

Poultney, C.S., Gutierrez, R.A., Katari, M.S., Gifford, M.L., Paley, W.B., Coruzzi, G.M.,et al. 2007. Sungear: interactive visualization and functional analysis of genomicdatasets. *Bioinformatics*, 23:259-261.

Prusinkiewicz, P. and Rolland-Lagan, A.G. 2006. Modeling plant morphogenesis. *Curr. Opin. Plant Biol.*, 9:83–88.

Rabinowicz, P.D., McCombie, W.R. and Martienssen, R.A. 2003. Gene enrichmentin plant genomic shotgun libraries. *Curr Opin Plant Biol* 6:150-156.

Rabinowicz, P.D., Schutz, K., Dedhia, N., Yordan, C., Parnell, L.D., Stein, L., et al. 2008. c01 BLBK130-*Coruzzi*, 16:42

Reinhardt, D., Frenz, M., Mandel, T. and Kuhlemeier, C. 2003. Microsurgical andlaser ablation analysis of interactions between the zones and layers of the tomatoshoot apical meristem. *Development*, 130:4073-4083.

Shannon, P., Markiel, A., Ozier, O., Baliga, N.S.,Wang, J.T., Ramage, D., et al. 2003. Cytoscape: a software environment for integrated models of biomolecular interactionnetworks. *Genome Res.*, 13:2498–2504.

Sharma, V.K., Carles, C. and Fletcher, J.C. 2003. Maintenance of stem cell populationsin plants. *Proc Natl Acad Sci USA* 100 (Suppl 1):11823-11829.

Snoep J.L. and Westerhoff, H.V., Alberghina, L. and Westerhoff, H.V. (Eds.) 2005. From isolation to integration, a systems biology approach for building the Silicon Cell. Systems Biology: Definitions and Perspectives. Springer-Verlag. p. 7.

Stelling, J., Sauer, U., Szallasi, Z., Doyle, F.J., III and Doyle, J. 2004. Robustness ofcellular functions. *Cell* 118:675-685.

Thimm, O., Blasing, O., Gibon, Y., Nagel, A., Meyer, S., Kruger, P., et al. 2004. MAPMAN: a user-driven tool to display genomics data sets onto diagrams of metabolicpathways and other biological processes. *Plant J.* 37:914-939.

Turing, A.M. 1952. The chemical basis of morphogenesis. *Philos Trans R Soc Lond B.Biol. Sci.* 237:37-72.

Wagner, A. and Fell, D.A. 2001. The small world inside large metabolic networks. *Proc Biol Sci* 268:1803-1810.

Wilkinson, M.D., Gessler, D., Farmer, A. and Stein, L. 2003. The BioMOBY project explores open-source, simple, extensible protocols for enabling biological database interoperability. *Proc Virt Conf Genom Bioinf,* 3:16–26.

Wilkinson, M.D. and Links, M. 2002. BioMOBY: an open source biological web services proposal. Brief Bioinform, 3:331–341.

Winter, D., Vinegar, B., Nahal, H., Ammar, R.,Wilson, G.V. and Provart, N.J. 2007. An'electronic fluorescent pictograph' browser for exploring and analyzing large-scalebiological data sets. PLoS ONE 2, e718.

Yuan, Y., SanMiguel, P.J. and Bennetzen, J.L. (2003) High-Cot sequence analysis of themaize genome. *Plant J.,* 34:249-255.

28

Bio-efficacy of Indigenous *Bacillus thuringiensis* LDC-9 Against Insect Pest and its Characterization through *In-silico* Approach

R. Shenbagarathai and A. Mahalakshmi

INTRODUCTION

Insects are the major group of arthropods and the most diverse group of animals across the Earth. Though some of them are beneficial to environment and to humans, many insects are considered as pests as they transmit diseases and damage the crops and forests. Chemical, biological, cultural and mechanical methods are being used to eradicate, control or manage these pests. The main strategy used for controlling pests has been the use of chemical pesticides because of their long residual action and wide spectrum toxicity (Casida and Quistad, 1998). However, the indiscriminate usage of chemical pesticides are toxic to mammals, birds and fish, also pollute the environment, besides the development of pest resistance and the emergence of secondary recalcitrant insect pest populations (Li *et al.,* 2007). These negative impacts prompted the change of strategies in managing the pests by resorting to ecofriendly and cost effective pest control strategies (Ferre *et al.,* 1991). The most promising strategy is the biological control method (Li *et al.,* 2007).

Biological control includes all biological material such as plant based products, microorganisms such as bacteria, fungi and viruses, natural enemies such as predators and parasites. Microbial Pest Control Agents (MPCAs), notably products of various *B.thuringiensis* (*B.t*) subspecies, are increasingly used in pest management programmes as *B.t* is regarded as safe due to its high specificity against target insects, high effectiveness, narrow host range, lesser residual efficacy, and is completely biodegradable (Peng *et al.,* 2007).

Bacillus thuringiensis is the Gram-positive soil bacterium. It produces δ-endotoxins during the sporulation phase of its growth cycle andserves as an important reservoir of Cry toxins and *cry* genes for production of biological insecticides. When insects ingest toxin crystals, the alkaline pH of their digestive tract causes the inactive protoxin to active toxin (Ferre *et al.,* 1991). It becomes inserted into the insect's gut cell membranes forming a pore resulting in swelling, cell lysis and eventually killing the insect. The combination of high activity on target insects with minimal non-target toxicity has made these toxins very attractive as biocontrol tools against insect speciesamong the orders Lepidoptera, Diptera, Coleoptera, Hymenoptera, Homoptera,Phthiraptera,Orthoptera, Mallophaga, Acari, Nemathelminthes, Platyhelminthes, Sarcomastigorphora and protozoan pathogens (Glare and O'Callaghan, 2000).

Approximately, more than 260 *cry* genes have been classified into large number of groups and subgroups based on their amino acid sequence homology .*B.t* is a gram-positive, spore-forming bacterium, capable of producing δ-endotoxins during the sporulation phase of its growth cycle (http://www.biols.susx.ac.uk/Home/Neil_Crickmore/Bt/). Numerous commercial *B.t* based bioproducts such as Dipel®, Javelin®, Thuricide®, Worm Attack®, Caterpillar Killer®, Bactospeine®, SOK-Bt®, VectoBac® and Teknar® have achieved moderate commercial success in developed countries .The high cost, poor stability and need for frequent reapplication deter their use in many developing countries (Gayathri *et al.,* 2004). Notwithstanding the variability of Cry proteins described up to now, it is still necessary to search for more toxins, sincea significant number of pests are not controlled with the available Cry proteins (Van Rie *et al.,*1991).

SCREENING

Screening programmes are expected to have an impact on the control of insects worldwide by increasing the diversity of toxins. Hence, Attempts are made to understand the distribution of *B.t* and its ecology .*B.t* isolates from Madurai region exhibited a great diversity in this study with *cry4* (50%) followed by *cry2* (45%). These results coincided with Ben-Dov *et al.*1997, wherein the specific-genes (*cry4)* were more frequently found and their distribution correlated with the distribution of dipteran insects. Contrastingly, earlier reports of Chak *et al.,* 1994; Bravo *et al.,* 1998; Martinez and Caballero, 2002 indicated relatively a low proportion of strains with *cry4*. Some genes such as *cry11* (29%),*cry10* (25%), *cyt1* (9%) and *cyt2* (7%) showed variation in the frequency of gene distribution. These results differed from earlier studies of Bravo *et al.* in 1998 who reported that 38% of the tropical strains were with *cry11* and *cyt* genes. Interestingly, the *cry1* gene was absent in this field collection although the standard *B.t* HD-1 confirmed the presence of the *cry1* gene by the PCR analysis. Moreover, the Phase Contrast Microscopic observation also revealed the presence of spherical paraspora inclusion. These results are different from the studies of Prabagaran *et al.* (2002)who reported the occurrence of the *cry1* genes from regions of India other than Madurai. It is important to mention that only a few of the strains in this collection harbored more than one *cry* gene. This result is different from the studies reported earlier (Ben-Dov *et al.*, 1997).

BIOASSAY

B.t isolates, which promoted 50% or more larval mortality, are considered as active strains according to Hossain *et al.*,(1997). *B.t* strains exhibited toxicity spectrum in the range of about 0-100% for dipteran larvae (*Culex quinquefasciatus*). Toxicity tests revealed that only 7% of the recovered *B.t* isolates were pathogenic (>50%) to dipteran larvae and 93 % of the *B.t* collection exhibited less than 50% toxicity to Dipteran larvae.The high proportion of the inactive strain observed in our study agreed with earlier reports (Ohba *et al.,* 2009). The vast variation observed in the toxicity of different strains may likely be due to the presence or absence of *cry* or *cyt* genes. Loss of plasmid from *B.t* strains is likely to make them avirulent/non-pathogenic/non-toxic (Balaraman, 2005). Interestingly, only one strain *B.t* LDC-9 exhibited toxicity (100%) to *Culex quinquefasciatus*.

A large number of isolates with dipteran toxicity are observed. The strains also differed in their toxicity (0-100%) to *Culex quinquefasciatus*. Martinez and Caballero (2002) suggested that this result might be explained by a variation in the level of gene expression, which can strongly influence the insect toxicity of the bacterial strain. The lack of toxicity against dipteran larvae is attributed to the low or non-expression of the active *cry* genes or genes detected are cryptic as has been previously reported (Masson *et al.*, 1998). Among 417 inter-positive isolates, only *B.t* LDC-9 showed interesting combinations of *cry* and *cyt* genes (*cry4A, cry4B, cry10Aor cry11A, cyt1, cyt2*) that exhibited high toxic activity to *Culex quinquefasciatus* and *Aedes aegypti*.This result is similar to the previous discovery of strains toxic to the larvae of some members of the genera Diptera (Johnson and Bishop, 1996). These data support the idea that although a great variability in *cry* genes coding for different mosquitocidal toxins exists in nature. One of the most effective combinations of proteins is that present in the *B. t* subsp. *israelensis* strain, containing *Cry4, Cry10, Cry11*, and Cyt toxins. Our results agreed with earlier reports that the isolates with spherical parasporal crystals are mosquito active (Ohba *et al.,* 2000).

The PCR products obtained were sub-jected to direct sequencing (Sanger *et al.*, 1977) and submitted to Genbank (Table 1). The current nomenclature, based solely on amino acid identity, allows closely related toxins to be ranked together.

Each new toxin is assigned a unique name incorporating four ranks. The toxin sequences obtained in this study is available at http://www.lifesci.sussex.ac.uk/home/Neil_Crickmore/Bt/

Table 1: Sequences used in this study

Accession Number	Gene	bp
AJ639659	ITS sequence	204
DQ078744	Wild type strain *cry4A*	434
DQ174290	Wild type strain *cry4A*	1489
DQ078743	Wild type strain *cry4B*	264
DQ167578	Wild type strain *cry10Aa*	328
DQ166531	Wild type strain *cry11A*	302
DQ166531	Wild type strain cyt1Aa	429
DQ200984	Wild type strain *cyt2B*	358
DQ910835	Mutant *cry4A*	435
DQ910833	Mutant *cry4B*	288
DQ910834	Mutant *cry10A*	326
DQ914923	Mutant *cry 11A*	320

LOCAL SIMILARITIES AND PAIRWISE SEQUENCE ALIGNMENT

Sequence similarity tools compare a query sequence to sequences in a database and return the best hits. It aims at finding sequences related to the query and, by similarity, at identifying properties of the query. Usually, query and database sequences may be amino acid or nucleotide sequences. Construction of an optimal alignment is the central problem in the entire field of biological sequence analysis. It is important to determine the degree of this similarity and to delineate similar segments with maximum precision.

BLAST ANALYSIS (HTTP://BLAST.NCBI.NLM.NIH.GOV/BLAST.CGI)

BLAST was used to input a nucleotide or protein sequence as a query against all (or a subset of) the public sequence databases, pasting the sequence into the textbox on one of the BLASTWeb pages. In this study, an optional Organism limit was employed to refine the query. M*egablast* (default), *discontiguous megablast* and *blastn* were selected for nucleotide similarity searches.*Blastp, PSI-BLAST and PHI-BLAST* are used for protein searches. The search is performed on the NCBI databases and servers, and the results are posted back to the person's browser in the chosen display format.

The report consists of three major sections: (1) the header, which contains information about the query sequence, the database searched (Fig.1A). On the Web, there is also a graphical overview (Fig.1B); (2) the one-line descriptions of each database sequence found to match the query sequence; these provide a quick overview for browsing (Fig.1C); (3) there may be more than one alignment for a database sequence it matches (Fig.1D).

Figure1: BLAST analysis

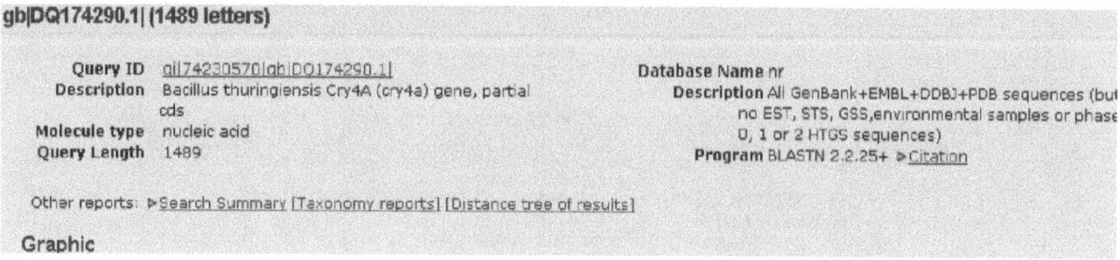

Fig.1A: Header

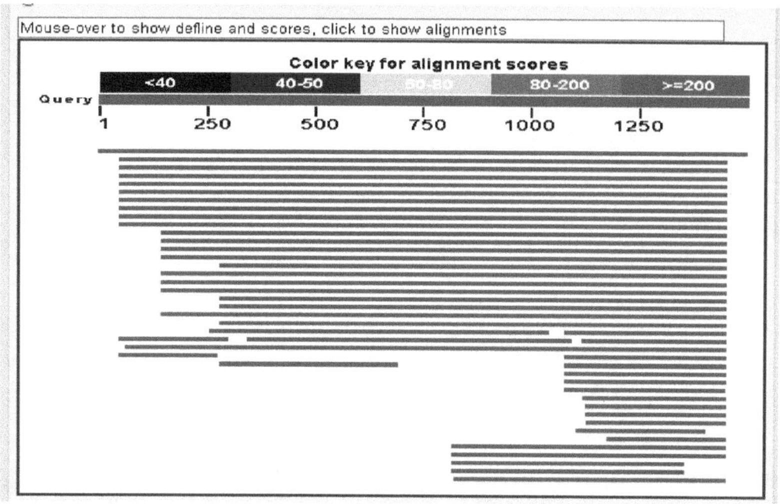

Fig.1B: Graphical overview

Sequences producing significant alignments:

Accession	Description	Max score	Total score	Query coverage	E value
DQ174290.1	Bacillus thuringiensis Cry4A (cry4a) gene, partial cds	2750	2750	100%	0.0
AL731825.1	Bacillus thuringiensis subsp. israelensis plasmid pBtoxis	2460	5826	93%	0.0
D00248.1	Bacillus thuringiensis israelensis plasmid gene for 130 kDa insecticida	2460	2460	93%	0.0
Y00423.1	Bacillus thuringiensis gene for 130 kDa delta-endotoxin	2455	2455	93%	0.0
EF424470.1	Bacillus thuringiensis isolate DAB-BT6 Cry4A gene, complete cds	2429	2429	93%	0.0
EF208904.1	Bacillus thuringiensis Cry4A (cry4A) gene, complete cds	2429	2429	93%	0.0
EF424472.1	Bacillus thuringiensis isolate DAB-BT4 Cry4A gene, complete cds	2418	2418	93%	0.0
EF424468.1	Bacillus thuringiensis isolate DAB-BT3 Cry4A gene, complete cds	2407	2407	93%	0.0
EF424471.1	Bacillus thuringiensis isolate DAB-BT2 Cry4A gene, complete cds	2396	2396	93%	0.0
EF424469.1	Bacillus thuringiensis isolate DAB-BT5 Cry4A gene, complete cds	2396	2396	93%	0.0
AY847707.1	Bacillus thuringiensis strain PBT602 130 kDa crystal protein (cry) ger	2054	2054	87%	0.0
X07423.1	Bacillus thuringiensis israelensis bt8 gene for 130 kDa crystal protein	2054	2054	87%	0.0
X07082.1	Bacillus thuringiensis gene for 130 kDa delta-endotoxin	2054	2054	87%	0.0
D00247.1	Bacillus thuringiensis israelensis plasmid gene for 130 kDa insecticida	2054	2054	87%	0.0
EF468627.1	Bacillus thuringiensis isolate DAB-Bt3 pesticidal crystal protein gene,	2028	2028	78%	0.0
EF468625.1	Bacillus thuringiensis isolate DAB-Bt1 pesticidal crystal protein gene,	2026	2026	87%	0.0
AY729887.1	Bacillus thuringiensis serovar israelensis delta-endotoxin (cry4BLB) g	2026	2026	87%	0.0
EF468626.1	Bacillus thuringiensis isolate DAB-Bt2 truncated pesticidal crystal prc	2021	2021	87%	0.0
EF468628.1	Bacillus thuringiensis isolate DAB-Bt4 pesticidal crystal protein gene,	1995	1995	78%	0.0
EF468630.1	Bacillus thuringiensis isolate DAB-Bt6 pesticidal crystal protein gene,	1989	1989	78%	0.0
M20242.1	B.thuringiensis mosquitocidal protein (CryD2) gene, complete cds	1984	1984	87%	0.0
EF468629.1	Bacillus thuringiensis isolate DAB-Bt5 pesticidal crystal protein gene,	1962	1962	78%	0.0

Fig.1C: One-line descriptions of each database sequence found to match the query sequence

```
>gb|DQ174290.1|  Bacillus thuringiensis Cry4A (cry4a) gene, partial cds
Length=1489

 Score = 2750 bits (1489),  Expect = 0.0
 Identities = 1489/1489 (100%), Gaps = 0/1489 (0%)
 Strand=Plus/Plus

Query  1    GTGTTCATGGACCTAGTGAAGTGTGTGTATCACTCAACGATACTTTCGGGTACTTAAATT  60
            ||||||||||||||||||||||||||||||||||||||||||||||||||||||||||||
Sbjct  1    GTGTTCATGGACCTAGTGAAGTGTGTGTATCACTCAACGATACTTTCGGGTACTTAAATT  60

Query  61   TTCTAACGAGTGAAATTTGCTCCAAATCAAAACATATCTCTTGTGTTTAATCGTTCGGAT  120
            ||||||||||||||||||||||||||||||||||||||||||||||||||||||||||||
Sbjct  61   TTCTAACGAGTGAAATTTGCTCCAAATCAAAACATATCTCTTGTGTTTAATCGTTCGGAT  120

Query  121  GTATATACAAACACAACAGTACTTATTGATAAAATTGAATTTCTGCCAATTACTCGTTCT  180
            ||||||||||||||||||||||||||||||||||||||||||||||||||||||||||||
Sbjct  121  GTATATACAAACACAACAGTACTTATTGATAAAATTGAATTTCTGCCAATTACTCGTTCT  180

Query  181  ATAAGAGAGGATACAGAGAAACAAAAATTAGAAACAGTACAACAAATAATTAATACATTT  240
            ||||||||||||||||||||||||||||||||||||||||||||||||||||||||||||
Sbjct  181  ATAAGAGAGGATACAGAGAAACAAAAATTAGAAACAGTACAACAAATAATTAATACATTT  240

Query  241  TATGCAAATCCTATAAAAAACACTTTACAATCAGAACTTACAGATTATGACATAGATCAA  300
            ||||||||||||||||||||||||||||||||||||||||||||||||||||||||||||
Sbjct  241  TATGCAAATCCTATAAAAAACACTTTACAATCAGAACTTACAGATTATGACATAGATCAA  300

Query  301  GCCGCAAATCTTGTGGAATGTATTTCTGAAGAATTATATCCAAAAGAAAAAATGCTGTTA  360
            ||||||||||||||||||||||||||||||||||||||||||||||||||||||||||||
Sbjct  301  GCCGCAAATCTTGTGGAATGTATTTCTGAAGAATTATATCCAAAAGAAAAAATGCTGTTA  360

Query  361  TTAGATGAAGTTAAAAATGCGAAACAACTTATTCAATCTCGAATGTACTTCACAACGGGG  420
            ||||||||||||||||||||||||||||||||||||||||||||||||||||||||||||
Sbjct  361  TTAGATGAAGTTAAAAATGCGAAACAACTTATTCAATCTCGAATGTACTTCACAACGGGG  420

Query  421  ATTTTGAATCCGCTACCTTGGTTGGACACAAGTGATAATATCACAATTCAGAAAATGATC  480
            ||||||||||||||||||||||||||||||||||||||||||||||||||||||||||||
Sbjct  421  ATTTTGAATCCGCTACCTTGGTTGGACACAAGTGATAATATCACAATTCAGAAAATGATC  480
```

Fig.1D: The alignments for each database sequence matched

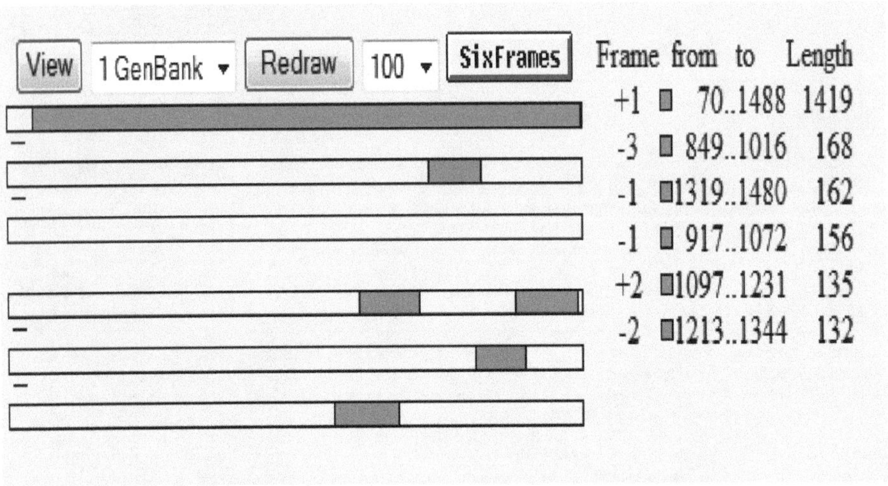

Fig. 2: Prediction of open reading frame using NCBI ORF Finder

PREDICTION OF PROTEIN-CODING REGIONS IN NUCLEOTIDE SEQUENCES

Nucleotide sequences are "translated" in order to generate all possible open reading frames (ORFs) i.e. sequence segments containing no termination codons. Translation is performed in all 6 phases, that is, 3 phases on the direct strand, and 3 ones on the complementary strand. 5'-proximal initiation (AUG) codons are The ORF Finder (Open Reading Frame Finder), a graphical analysis tool which finds all open reading frames of a selectable minimum size in a user's sequence. This tool was used to identify all open reading frames using the bacterial genetic codes. The frame which contained a long uninterrupted stretch of sequence was selected as the correct frame (Fig.2).The deduced amino acid sequence was searched against the sequence database using the WWW BLAST server. The Blast results confirm the presence of Cry protein and the predicted frame as the correct frame. http://www.ncbi.nlm.nih.gov/gorf/gorf.html

MULTIPLE SEQUENCE ALIGNMENT:

The simultaneous alignment of many nucleotide or amino acid sequences is now an essential tool in molecular biology. Multiple alignments are used to find diagnostic patterns to characterise protein families; to detect or demonstrate

```
              841                                                        900
c4a           KLKPYTRYLVRGFVGSSKDVELVVSRYGEEIDAIMNVPADLNYLYPSTFDCEGSNRCETS
Q7AL67_BAC    KLKPYTRYLVRGFVGSSKDVELVVSRYGEEIDAIMNVPADLNYLYPSTFDCEGSNRCETS
CR4AA_BACT    KLKPYTRYLVRGFVGSSKDVELVVSRYGEEIDAIMNVPADLNYLYPSTFDCEGSNRCETS
Q7AL72_BAC    KLKPYTRYLVRGFVGSSKDVELVVSRYGEEIDAIMNVPADLNYLYPSTFDCEGSNRCETS
Q29YB8_BAC    KLKPYTRYLVRGFVGSSKDVELVVSRYGEEIDAIMNVPADLNYLYPSTFDCEGSNRCETS
CR4BA_BACT    KLKPYTRYLVRGFVGSSKDVELVVSRYGEEIDAIMNVPADLNYLYPSTFDCEGSNRCETS
              ************************************************************
              901                                                        960
c4a           AVPANIGNTSDMLYSCQYDTGKKHVVCQDSHQFSFTIDTGALDTNENIGVWVMFKISSPD
Q7AL67_BAC    AVPANIGNTSDMLYSCQYDTGKKHVVCQDSHQFSFTIDTGALDTNENIGVWVMFKISSPD
CR4AA_BACT    AVPANIGNTSDMLYSCQYDTGKKHVVCQDSHQFSFTIDTGALDTNENIGVWVMFKISSPD
Q7AL72_BAC    AVPANIGNTSDMLYSCQYDTGKKHVVCQDSHQFSFTIDTGALDTNENIGVWVMFKISSPD
Q29YB8_BAC    AVPANIGNTSDMLYSCQYDTGKKHVVCQDSHQFSFTIDTGALDTNENIGVWVMFKISSPD
CR4BA_BACT    AVPANIGNTSDMLYSCQYDTGKKHVVCQDSHQFSFTIDTGALDTNENIGVWVMFKISSPD
              ************************************************************
              961                                                       1020
c4a           GYASLDNLEVIEEGAIDGEALSRVKHMEKKWNDQMEAKRLETQQAYDVAKQAIDALFTNV
Q7AL67_BAC    GYASLDNLEVIEEGPIDGEALSRVKHMEKKWNDQMEAKRSETQQAYDVAKQAIDALFTNV
CR4AA_BACT    GYASLDNLEVIEEGPIDGEALSRVKHMEKKWNDQMEAKRSETQQAYDVAKQAIDALFTNV
Q7AL72_BAC    GYASLDNLEVIEEGPIDGEALSRVKHMEKKWNDQMEAKRSETQQAYDVAKQAIDALFTNV
Q29YB8_BAC    GYASLDNLEVIEEGPIDGEALSRVKHMEKKWNDQMEAKRSETQQAYDVAKQAIDALFTNV
CR4BA_BACT    GYASLDNLEVIEEGPIDGEALSRVKHMEKKWNDQMEAKRSETQQAYDVAKQAIDALFTNV
              **************.************************ ********************
              1021                                                      1080
c4a           QDEALQFDTTLAQIQYAEYLVQSIPYVYNDWLSDVPGMNYDIYVELDARVAQARYVYDTR
Q7AL67_BAC    QDEALQFDTTLAQIQYAEYLVQSIPYVYNDWLSDVPGMNYDIYVELDARVAQARYLYDTR
CR4AA_BACT    QDEALQFDTTLAQIQYAEYLVQSIPYVYNDWLSDVPGMNYDIYVELDARVAQARYLYDTR
Q7AL72_BAC    QDEALQFDTTLAQIQYAEYLVQSIPYVYNDWLSDVPGMNYDIYVELDARVAQARYLYDTR
Q29YB8_BAC    QDEALQFDTTLAQIQYAEYLVQSIPYVYNDWLSDVPGMNYDIYVELDARVAQARYLYDTR
CR4BA_BACT    QDEALQFDTTLAQIQYAEYLVQSIPYVYNDWLSDVPGMNYDIYVELDARVAQARYLYDTR
              *******************************************************:****
              1081                                                      1140
```

Figure 3: Conserved domains observed in *B.thuringiensis* LDC-9 Cry4A protein homology between new sequences and existing families of sequences; to help predict the secondary and tertiary structures of new sequences and to suggest oligonucleotide primers for PCR, as an essential prelude to molecular evolutionary analysis. Multiple sequence alignment results (EMBL_Align database: ALIGN_000987) revealed the presence of a conserved endotoxin C-domain in Cry4A protein sequence of *B.t* LDC-9 that is in concor--dance with other Cry4A protein sequences (Fig.3). Multiple sequence alignment of all Cry proteins are submitted in EMBL-Align: a new public nucleotide and amino acid multiple sequence alignment database (Table 2).

Table 2: Multiple sequence alignment of *Bacillus thuringiensis* Cry proteins submitted in EMBL-Align: a new public nucleotide and amino acid multiple sequence alignment database.

EMBL-Align:ALIGN_001163	Alignment of Bacillus thuringiensis Cry10Aa protein sequence (P09662) with other experimentally determined Cry structures
EMBL-Align:ALIGN_001164	Alignment of Bacillus thuringiensis LDC-9 with other Cry sequences whose structure has been determined experimentally
EMBL-Align:ALIGN_001165	Alignment of Cry10Aa protein from Bacillus thuringiensis israelensis and B. thuringiensis LDC-9 with N-terminal consensus sequence of the endotoxin fold family
EMBL-Align:ALIGN_001186	Sequence alignment of Bacillus thuringiensis LDC-9 Cyt1 and Cyt2 proteins with existing Cyt proteins
EMBL-Align:ALIGN_001188	Alignment of cytolytic proteins with the conserved domain
EMBL-Align:ALIGN_001195	Alignemnt of cytolytic proteins with the conserved domain
EMBL-Align:ALIGN_001196	Alignment of cytolytic proteins from bacillus thuringiensis
EMBL-Align:ALIGN_001198	Alignment of Cry10A protein sequences with N-terminal endotoxin domain
EMBL-Align:ALIGN_001200	Alignment of complete Cry10 protein seq with N-terminal endotoxin domain
EMBL-Align:ALIGN_001202	Alignment of Bacillus thuringiensis LDC-9 Cry10A protein sequences with other protein sequences in the SwissProt Database
EMBL-Align:ALIGN_001204	Alignment of Bacillus thuringiensis LDC-9 Cry11A with other Cry11A sequences
EMBL-Align:ALIGN_001205	Alignment of Bacillus thuringiensis cytolytic protein sequences (complete)

EMBL-Align	DESCRIPTION
EMBL-Align:ALIGN_000841	Alignment of cry2Ab genes
EMBL-Align:ALIGN_000985	Alignment of ITS1 from B.thuringiensis LDC-9 and other isolates
EMBL-Align:ALIGN_000986	Alignment of partial 16S rRNA gene from indigenous Bacillus sp.
EMBL-Align:ALIGN_000987	Alignment of partial Cry4a gene from indigenous B.thuringiensis
EMBL-Align:ALIGN_001020	Alignment of pesticidial crystal protein cry10AA from Bacillus thuringiensis
EMBL-Align:ALIGN_001022	Alignment of Cry10Aa protein from Bacillus thuringiensis with other significant sequences
EMBL-Align:ALIGN_001023	Alignment of Cry11Aa from Bacillus thuringiensis with other Cry11 sequences
EMBL-Align:ALIGN_001024	Alignment of Cry4B protein from Bacillus thuringiensis with other significantly similar proteins
EMBL-Align:ALIGN_001025	Alignment of Bacillus thuringiensis Cry4B protein with other Cry4 proteins
EMBL-Align:ALIGN_001026	Alignment of Cyt1Aa protein from Bacillus thuringiensis with other cytolytic proteins.
EMBL-Align:ALIGN_001027	Alignment of Cyt2Ba protein from Bacillus thuringiensis with other cytolytic proteins.
EMBL-Align:ALIGN_001028	Multiple sequence alignment of Cyt2B protein from Bacillus thuringiensis with other Cyt2B protein sequences
EMBL-Align:ALIGN_001029	Alignment of Cry4A protein from Bacillus thuringiensis with other Cry4 proteins
EMBL-Align:ALIGN_001030	Alignment of 16S rRNA gene from Bacillus thuringiensis
EMBL-Align:ALIGN_001148	Alignment of B. thuringiensis LDC-9 Cyt1, B. thuringiensis cytolytic proteins with cytolytic domain observed from conserved domain search using BLAST
EMBL-Align:ALIGN_001149	Alignment of cytolytic proteins from B. thuringiensis species
EMBL-Align:ALIGN_001161	Alignment of P09662 with Cry proteins whose experimental structures are available.
EMBL-Align:ALIGN_001162	Alignment of Bacillus thuringiensis Cry10Aa protein sequence with a C-terminal domain

CONSTRUCTION OF PHYLOGENETIC TREE:

Construction of phylogenetic tree is, in a sense, the final goal of analysis of biological sequences. Phylogenetic relationships among proteins in different organisms may be inferred from the protein sequences. The basic idea is that more mutations are required to change one protein sequence into the other; the more unrelated the sequences and the lower the probability that they share a recent common ancestor sequence. A tree structure of the proteins can be used to describe the evolutionary relationship among a family of proteins (Fig.4).

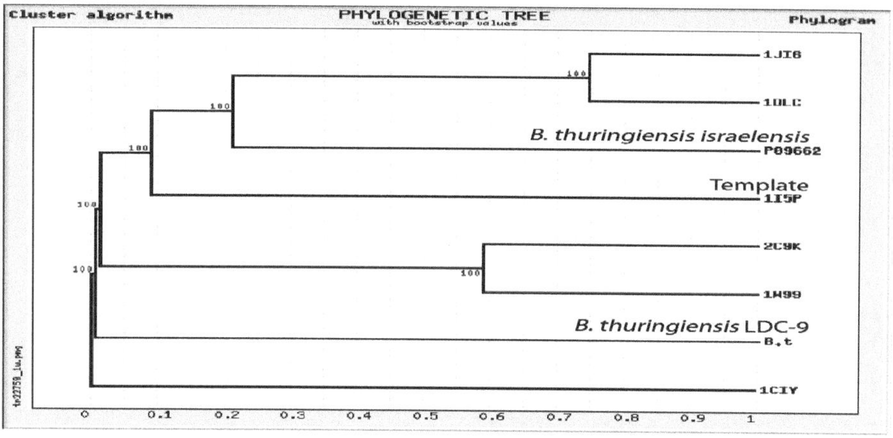

Fig. 4: Phylogenetic tree showing evolutionary relationships of Cry protein sequences.

CRY10A MODELLING

Since the structural information of Cry10 δ-endotoxins is unavailable, this work focuses on proposing a model for the structure of the Cry10Aa δ-endotoxin based on the hypotheses of structural similarity with existing Cry proteins, which are experi-mentally determined. This model provides information for the design of muta-genesis experiments aimed to elucidate the mechanism of action of the Cry10 family of toxins. Comparative or homology protein structure modeling builds a three-dimensional model for a protein of unknown structure based on one or more related proteins of known structure as reported by Sanchez *et al.,* (2000). Comparative modeling usually starts by searching the PDB of known protein structures using the target sequence as the query. This search is done by comparing the target sequence with the sequence of each of the structures in the database.

MODEL STRUCTURE

The modeled Cry10 structure (Fig. 5) confirms the domain topology (three domains) observed in the majority of *B. t* toxins characteristic of the delta-endotoxin family as previously reported (Fig. 6) for Cry1Aa (Grochulski *et al.,* 1995), Cry3Aa

(Li *et al.*,1991), Cry3Bb1 (Galitsky *et al.,* 2001), Cry2Aa (Morse *et al.*,2001), Cry4Ba (Boonserm *et al.*,2005) and Cry4Aa (Boonserm *et al.*,2006). Reports of similar 3-domain structure despite the low sequence similarity were observed in the structures of activated toxins such as Cry1Aa (PDB : 1CIY) (Grochulski *et al.,* 1995), Cry3Aa (PDB : 1DLC) (Li *et al.*,1991), Cry3Bb1 (PDB : 1JI6) (Galitsky *et al.*,2001) and Cry2Aa (PDB : 1I5P) (Morse *et al.*,2001). Superpositioning macromolecular structures is an essential tool in structural bioinformatics and is used routinely in the fields of NMR, X-ray crystallography, protein folding, molecular dynamics and structural evolution.

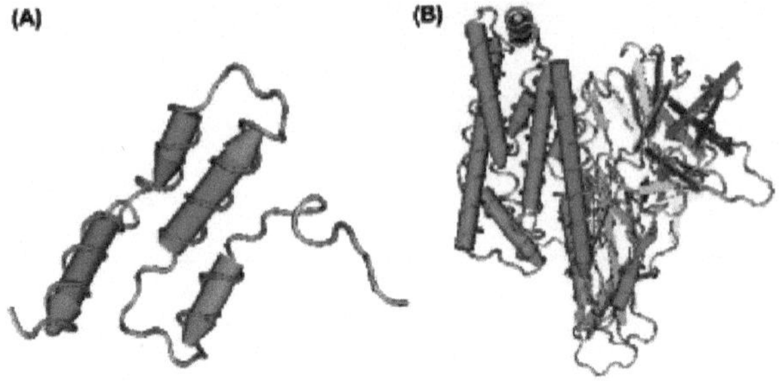

Fig. 5: Theoretical predicted model of Cry10A protein. **A.** *B.thuringiensis* LDC-9 **B.** *B.thuringiensis israelensis*

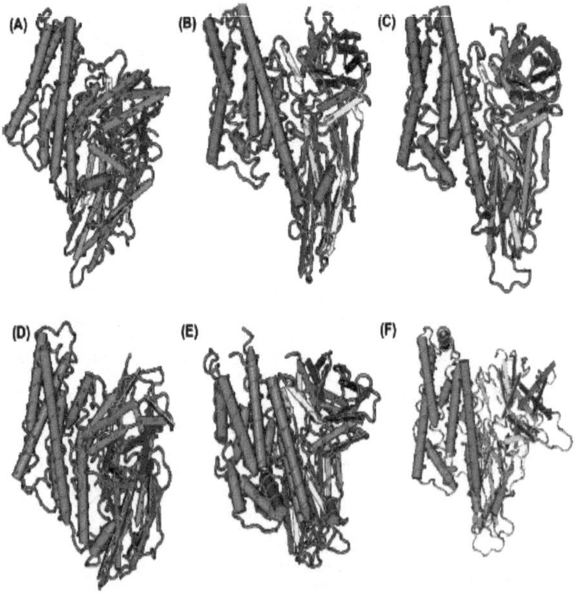

Fig.6: Comparative account of the three domain experimentally determined proteins with theoretically determined *B. thuringiensis israelensis* (P09662) model. Domains II and III pack on the same side of domain, along helix $\alpha 7$. **A.** 1I5P (Cry2Aa), **B.** 1JI6 (Cry3Bb), **C.** 1CIY(Cry1Aa), **D.** 1DLC(Cry3Aa), **E.** 2C9K(Cry4Aa), **F.** Theoretically determined 3D structure of *B. thuringiensis israelensis* Cry10A (P09662).

Superpositioning allows the comparison of structures by fitting their atomic coordinates to each other as closely as possible. The valid interpretation of the superposition relies upon the quality of the estimated orientations of the molecules and thus reliable, robust superpositioning tools are a critical component of structural analysis and comparison. Similarity is reflected in the modeled three-dimensionalstructures as the corresponding domains can virtually be superimposed (Fig.7).The outputs of a structural alignment are a superposition of the atomic coordinate sets and a minimal root mean square deviation (RMSD) between the structures. The RMSD of two aligned structures indicates their divergence from one another. It has been recognized, however, that the atomic RMSD ceases to be a reliable measure when it reaches the 3-7 Å range (Alexandrov and Go, 1994).

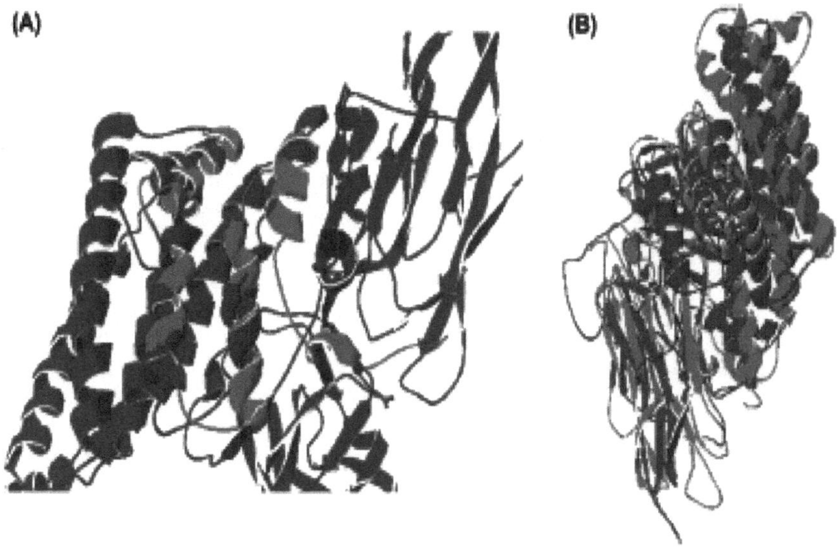

Fig. 7 **A.** Superposition of the structural domains of *B. thuringiensis* LDC-9 Cry10Aa wild type and Cry2Aa template from PDB (1I5P). Red color indicates *B. thuringiensis* LDC-9 Cry10Aa, blue color indicates Cry2Aa (1I5P). **B.** Superposition of the structural domains of *B. thuringiensis israelensis* (P09662) and Cry2Aa template from PDB (1I5P). Red color indicates *B. thuringiensis israelensis* (P09662), Blue color indicates Cry2Aa (1I5P). RMS DEVIATION-1.57A° Superposition was done only in the aligned regions.

RAMACHANDRAN PLOT

It is one of the simplest and most sensitive means for assessing the quality of a protein model in the absence of experimental data (Kleywegt and Jones, 1996). The *B. t* LDC-9 and the *B. t* subsp. *israelensis* indicated that 82.8% and 87.7% of the residues of the model respectively have f and y angles in the most allowed regions of the Ramachandran map indicating the model is acceptable (Fig. 8). Deviations of the atomic volumes from the standard values, evaluated as the volume

Z-scores, are used to assess the quality of the protein crystal structures. The volume-based structure validation procedures are implemented in the program PROVE (PROtein Volume Evaluation), which is accessible through the World Wide Web. The Z-sores obtained in this study are significant as the Z-score above 7.0 and below 7.0 represents the occurrence of many errors in structure in terms of the packing of the buried atoms (Pontius *et al.,* 1996). The models conform to the stereochemistry derived from small molecules with the RMSD of 0.005 Å in bond lengths, 1.797 Å in bond angles, 27.176 Å in dihedral angles and 1.068 Å in improper rotations.

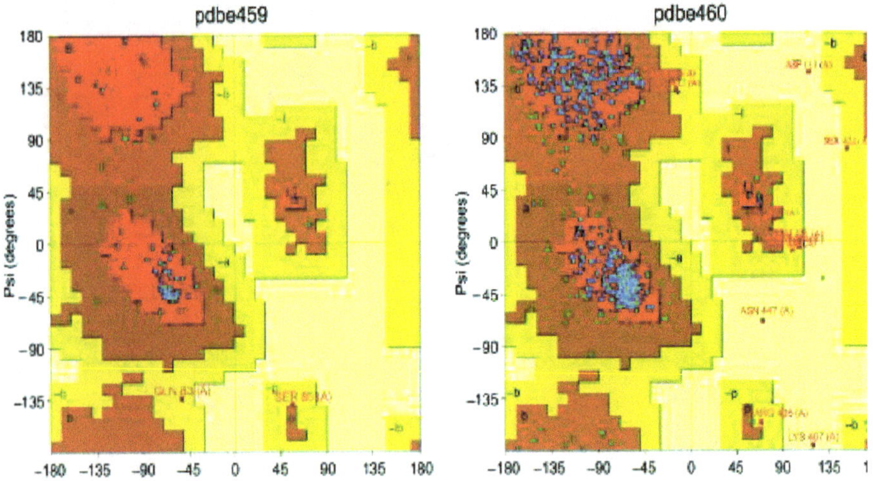

Fig. 8: Ramachandran plot for the models predicted. A. *B.thuringiensis LDC-9 Cry10A*, B. *B.thuringiensis israelensis Cry10A(P09662)*. The *darkest* areas shown in *red*) correspond to the *"core"* regions representing the most *favourable* combinations of *phi-psi* values. The different regions on the *Ramachandran plot* are as described in Morris et al., (1992). The regions are labelled as follows: *A,* - Core alpha, *a* - Allowed alpha, *~a* - Generous alpha, B - Core beta, *a* - Allowed beta, *~a* - Generous beta, *L* - Core left-handed alpha, *l* - Allowed left-handed alpha, *~l* - Generous left-handed alpha, *p* - Allowed epsilon, *~p* - Generous epsilon.

Hofte and Whiteley (1989) drew attention to the five blocks of amino acids conserved among most of the Cry toxins then known.The complete amino acid sequence alignment of the Cry proteins in the Cry10 data set reveals the same five tracts, or conserved blocks in *B. t.* subsp. *israelensis* Cry10Aa (UniProt: P09662, 675 aa) . Comparison of the carboxyl-terminalhalves of sequences suggests thepresence of three additional blocks lying outside the active toxiccore. The C-terminal extension does not have a role in the activity (it is digested by proteases in the insect gut) but it is believed to have a role in the formation of the crystal (Schnepf *et al.*, 1998).

Domain I, a bundle of eight antiparallel helices., The long hydrophobic and amphipathic helices of domain I is involved either in the initial interactions between the toxin and the epithelial gut lining of the insect or in a subsequent step in the disruption of the membrane as previously reported (Thorne *et al.*,1986). Several studies have shown that domain I is responsible for monomer-monomer interaction and ion channel formation (Grochulski *et al.*,1995). The comparison of the Cry domain I helices in the model of the *B. t.* subsp. *israelensis* with other known classes of amphipathic helices suggests that the helices α7-α5 are known to undergo a substantial conformation rearrangement of tertiary structure in order to form ion channels in the membrane as previously reported (Li *et al.*,1991) while α5 is critical for aggregation (Aronson *et al.*,1999).

Domain II, comprised of three antiparallel β-sheets is associated with receptor recognition and binding (Schnepf *et al.*,1998). The specificity determining regions have been mapped primarily to domain II by the creation of reciprocal hybrid genes between closely related toxins of differing specificities (Smith and Ellar, 1994).

Two twisted antiparallel β-sheets forming a β-sandwich was observed in domain III. The role of this domain III is unclear but it seems important for fine-tuning insect specificity, protein stability and pore function (Rausell *et al.*,2004).

The conserved blocks observed are important for toxin stability within the insect gut, probably by contacts with neighbor helices (Fig. 9). These results are supported by Valdez *et al.*(2001) with mutagenesis studies. Conserved regions of the protein

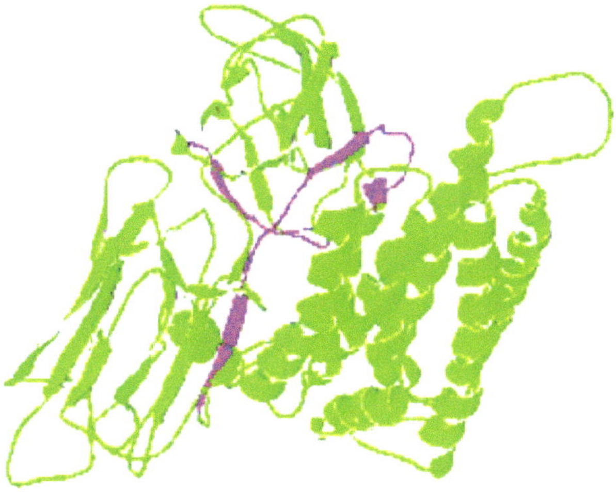

Fig. 9: Interdomain contacts of *B.thuringiensis israelensis* Cry10A(P09662). Pinkcolor represents interdomain contacts, Green color represents the other domains

in data set reveal the same five tracts, or conserved blocks in the model (Fig.10) as previously reported by Hofte and Whiteley (1989). The location of these conserved regions as either residues in the core of individual domains or residues involved in interdomain contacts within the framework of the CryIII toxin as discussed earlier by Li *et al.* (1991).

Fig. 10: View of the *B. thuringiensis israelensis* Cry10Aa (P09662) molecule with the conserved sequence blocks marked in different colors. The Cá trace is drawn in thin black lines while the secondary structure elements in the conserved regions are highlighted. Segment, 1. red; segment 2, Yellow; segment 3, green; segment 4, cyan; segment 5, magenta

The available chimeric scanning mutagenesis data (Liang and Dean, 1994) define a candidate toxin-receptor binding surface on Cry2Aa that is comprised of a distribution of hydrophobic residues. The modeled structure of the *B. t* subsp. *israelensis* (PMDB id: PM0077935) has similar distribution of hydrophobic residues (Val521-Asp527) from b12a (522-527), Lys707-Ile709 from the b5-b6 loop and Ser773-Ile775 from the b7-b8 loop and is a part of the hydrophobic core. These residues are expected to destabilize the structure-interaction and are responsible for activity.

Electrostatic interactions play a central role in protein structure and function (Bashford, 2004). Electrostatic potentials are visualized by color-coding the molecular surface according to potential values (e.g., red for negative potentials and blue for positive potentials while a linear interpolation is used to color surface potentials between the two selected values (Fig.11). The initial receptor binding process was suggested to be mediated by hydrophobic and electrostatic interactions

(Schnepf *et al.,* 1998). Long-range electrostatic interactions result in an orientation of domain I with helices á7 and á5 approximately perpendicular to the lipid bilayer, as was proposed previously (Parker *etal.*,1989).

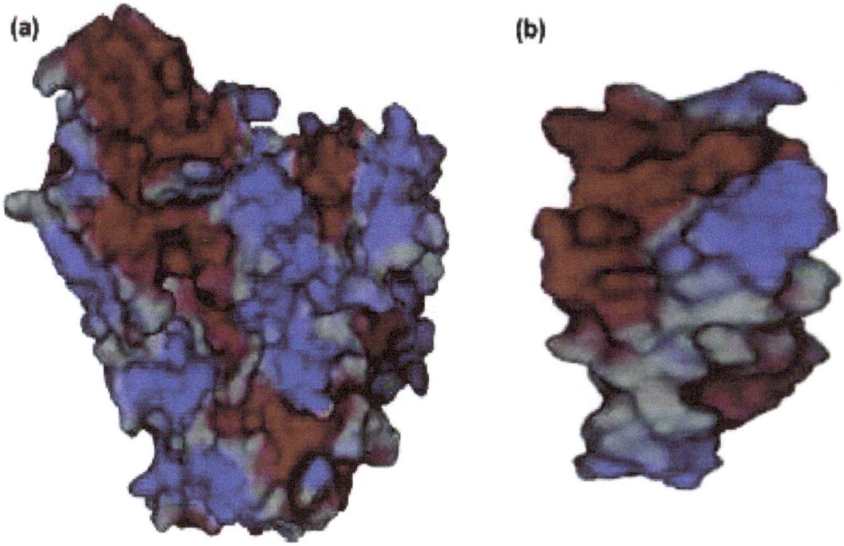

Fig. 11: Electrostatic potential surface of Cry10A **(a)** using Pymol. Red and blue represents positive and negative values respectively. a. *B. thuringiensis israelensis* (P09662), **(b)** *B. thuringiensis* LDC-9.

CYT1A MODELING

The conserved aromatic residues noted in the multiple sequence alignment of the Cytolytic family of proteins indicated an important role in cytolytic activity. This result is in concordance with earlier *in vivo* expression studies indicating the conserved aromatic residue at position 187 and tryptophan residues at positions 132 and 184 in Cyt2 are critical and essential to maintain the structure and functions of the protein (Promdonkoy *et al*., 2004). The prediction of the tryptic processing sites occurred after Arg28 in the N terminus of the Cyt1 family of proteins, map towards the C-terminus of b1, revealed by sequence alignment (Fig.12). Similar results for Cytproteins were previously reported (Al-yahyaee and Ellar, 1995). Earlier reports indicated differences in hemolytic or mosquitocidal activities between Cyt proteins, despite their high level of structural similarity, which might be due to the difference in processing (Li *et al.,* 1996). The processing cleaves the intertwined N-terminal arm, including the b1 responsible for the dimerisation, thereby releasing the active toxin as a monomer.

```
              61                                                         120
B.t           ISNYILQAIMLANAFQNALVPTSTDFG-DALRFSMPHGLEIANTITPMGAVVSYVDQNVT
Q38PK3_BAC    ISNYILQAIMLANAFQNALVPTSTDFG-DALRFSMPHGLEIANTITPMGAVVSYVDQNVT
Q29Y56_BAC    NPNYILQAIMLANAFQNALVPTSTDFG-DALRFSMPHGLEIANTITPMGAVVSYVDQNVT
Q7AL78_BAC    NPNYILQAIMLANAFQNALVPTSTDFG-DALRFSMPHGLEIANTITPMGAVVSYVDQNVT
Q2PTA3_BAC    NPNYILQAIMLANAFQNALVPTSTDFG-DALRFSMPHGLEIANTITPMGAVVSYVDQNVT
CT1AABACT     NPNYILQAIMLANAFQNALVPTSTDFG-DALRFSMPHGLEIANTITPMGAVVSYVDQNVT
CT1AA_BACT    NPNYILQAIMLANAFQNALVPTSTDFG-DALRFSMAHGLEIANTITPMGAVVSYVDQNVT
CT1AB_BACT    NTNYLLQAIMLANAFQKALVPTSTEFAEDALQFSMTHGLEVANTISPPGAVVQYVDQNVS
CT1BA_BACT    NPNYVLQAIQLAAAFQDALVPTETEFG-EAIRFSMPHGLEVAKTIQPHGAVVAYTDQTLS
Q93M48_BAC    APQYVNQALTLAKYFQGAIDGST-------LRFDFEKALQIANDI-PQAAVVNTLNQTVQ
Q8VNK0_BAC    EPQYIAQALRLTNTFQGAIDPLT-------LNFNFEKALQIANGL-PNAGVTGTLNQSVI
Q7AL73_BAC    EPQYIAQAIRLTNTFQGAIDPLT-------LNFNFEKALQIANGL-PNAGVTGTINQSVI
CT2BA_BACT    EPQYIAQAIRLTNTFQGAIDPLT-------LNFNFEKALQIANGL-PNAGVTGTINQSVI
Q71KW8_BAC    QPQYINQALHLANAFQGAIDPLN-------LNFNFEKALQIANGI-PNSAIVKTLNQSVI
CT2AA_BACT    QPQYINQALHLANAFQGAIDPLN-------LNFNFEKALQIANGI-PNSAIVKTLNQSVI
Q9L878_BAC    EPQYIAQAIRLTNTFQGAIDPLT-------LNFNFEKALQIANGL-PNAGVTGTIIQSVI
CT2BB_BACT    EPQYVAQALRLTNTFQAAIDPLT-------LNFDFEKALQIANGL-PNAGITGTLNQSVI
O30895_BAC    ------------NTFQGAIDPLT-------LNFNFEKALQIANGL-PNAGVTGTINQSVI
Q9R9C0_BAC    ------------NTFQGAIDPLT-------LNFNFEKALQIANGL-PNAGVTGTINQSVI
Q7BVD4_BAC    ------------NTFQGAIDPLT-------LNFNFEKALQIANGL-PNAGVTGTINQSVI
O52298_BAC    ------------NTFQGAIDPLT-------LNFNFEKALQIANGL-PNAGVTGTINQSVI
O30969_BAC    ------------NTFQGAIDPLT-------LNFNFEKALQIANGL-PNAGVTGTINQSVI
                             ** *:          :.*.: *.*::*: : * ..:.    *.:
```


```
              61                                                         120
B.t           ISNYILQAIMLANAFQNALVPTSTDFG-DALRFSMPHGLEIANTITPMGAVVSYVDQNVT
Q38PK3_BAC    ISNYILQAIMLANAFQNALVPTSTDFG-DALRFSMPHGLEIANTITPMGAVVSYVDQNVT
Q29Y56_BAC    NPNYILQAIMLANAFQNALVPTSTDFG-DALRFSMPHGLEIANTITPMGAVVSYVDQNVT
Q7AL78_BAC    NPNYILQAIMLANAFQNALVPTSTDFG-DALRFSMPHGLEIANTITPMGAVVSYVDQNVT
Q2PTA3_BAC    NPNYILQAIMLANAFQNALVPTSTDFG-DALRFSMPHGLEIANTITPMGAVVSYVDQNVT
CT1AABACT     NPNYILQAIMLANAFQNALVPTSTDFG-DALRFSMPHGLEIANTITPMGAVVSYVDQNVT
CT1AA_BACT    NPNYILQAIMLANAFQNALVPTSTDFG-DALRFSMAHGLEIANTITPMGAVVSYVDQNVT
CT1AB_BACT    NTNYLLQAIMLANAFQKALVPTSTEFAEDALQFSMTHGLEVANTISPPGAVVQYVDQNVS
CT1BA_BACT    NPNYVLQAIQLAAAFQDALVPTETEFG-EAIRFSMPHGLEVAKTIQPHGAVVAYTDQTLS
Q93M48_BAC    APQYVNQALTLAKYFQGAIDGST-------LRFDFEKALQIANDI-PQAAVVNTLNQTVQ
Q8VNK0_BAC    EPQYIAQALRLTNTFQGAIDPLT-------LNFNFEKALQIANGL-PNAGVTGTLNQSVI
Q7AL73_BAC    EPQYIAQAIRLTNTFQGAIDPLT-------LNFNFEKALQIANGL-PNAGVTGTINQSVI
CT2BA_BACT    EPQYIAQAIRLTNTFQGAIDPLT-------LNFNFEKALQIANGL-PNAGVTGTINQSVI
Q71KW8_BAC    QPQYINQALHLANAFQGAIDPLN-------LNFNFEKALQIANGI-PNSAIVKTLNQSVI
CT2AA_BACT    QPQYINQALHLANAFQGAIDPLN-------LNFNFEKALQIANGI-PNSAIVKTLNQSVI
Q9L878_BAC    EPQYIAQAIRLTNTFQGAIDPLT-------LNFNFEKALQIANGL-PNAGVTGTIIQSVI
CT2BB_BACT    EPQYVAQALRLTNTFQAAIDPLT-------LNFDFEKALQIANGL-PNAGITGTLNQSVI
O30895_BAC    ------------NTFQGAIDPLT-------LNFNFEKALQIANGL-PNAGVTGTINQSVI
Q9R9C0_BAC    ------------NTFQGAIDPLT-------LNFNFEKALQIANGL-PNAGVTGTINQSVI
Q7BVD4_BAC    ------------NTFQGAIDPLT-------LNFNFEKALQIANGL-PNAGVTGTINQSVI
O52298_BAC    ------------NTFQGAIDPLT-------LNFNFEKALQIANGL-PNAGVTGTINQSVI
O30969_BAC    ------------NTFQGAIDPLT-------LNFNFEKALQIANGL-PNAGVTGTINQSVI
                             ** *:          :.*.: *.*::*: : * ..:.    *.:
```


Fig. 12: Conserved domains observed in *B.thuringiensis* LDC-9 cytolytic proteins

The cytolytic toxin, when solubilised in the alkaline midgut, is released as a dimeric protoxin. The proteolytic activation of the protoxin at both N and C-termini is similar to earlier reports (Al-yahyaee and Ellar, 1995), a monomeric toxin is obtained that acts on the midgut epithelial cell membrane. Transmembrane prediction is considered significant and with high probability only when a region of the 18-22 amino acids with an average polarity value greater than 1.6 as reported earlier (Kyte and Doolittle, 1982).

The architecture and uncommon topology of the homology models predicted in this study (Fig. 13) agreed with the X-ray data of Cry2A structure previously reported (Li *et al.,* 1996). Theoretical screening of the model molecular surface also highlighted the potential importance of a similar hydrophobic patch in all the predicted models in comparison with the template (Li *et al.,* 1996). The *B. t* subsp. *israelensis* Cyt1A indicated that 99.5% of the residues of the model respectively have f and y angles in the most allowed regions of the Ramachandran map indicating the model is acceptable (Fig. 14).The regions of the helices, which are packed against the b-sheets, are hydrophobic , the long strand in the b-sheet showed an amphiphilic character, while the regions exposed to molecular surface are polar. These results are in agreement with earlier studies of Li *et al.* (1996). Further experimental evidence of the role of this hydrophobic surface came from mutagenesis studies, as substitutions of a few residues that were part of the hydrophobic cluster reduced dramatically the cytolytic activity (Ward and Ellar, 1986).

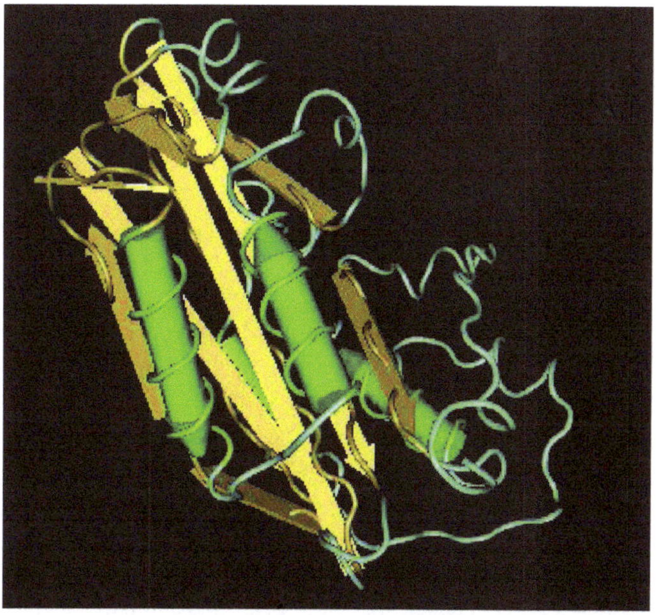

Fig. 13: Predicted theoretical model of *B.thuringiensis israelensis* Cyt1A

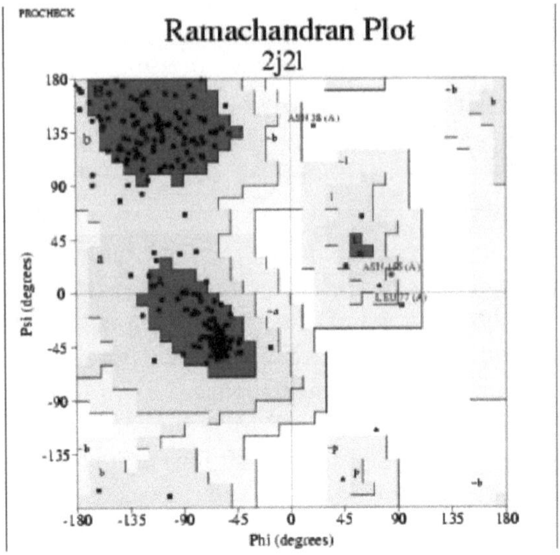

Fig. 14: Ramachandran plot for the models predicted. *B.thuringiensis israelensis Cyt1A*. The *darkest* areas shown in *red*) correspond to the *"core"* regions representing the most *favourable* combinations of *phi-psi* values. The different regions on the *Ramachandran plot* are as described in Morris *et al*., (1992). The regions are labelled as follows: *A*, - Core alpha, *a* - Allowed alpha, *~a* - Generous alpha, B - Core beta, *b* - Allowed beta, *~b* - Generous beta, *L* - Core left-handed alpha, *l* - Allowed left-handed alpha, *~l* - Generous left-handed alpha, *p* - Allowed epsilon, *~p* - Generous epsilon

PDB AND PMDB DATABASES

PMDB Protein Model Database, which collects three dimensional protein models obtained by structure prediction methods. The models predicted in this study are deposited in PDB and PMDB databases. The models are available at http://www.pdb.org/pdb/search/searchModels.do, http://mi.caspur.it/PMDB/.

CONCLUSION

In conclusion, the insilico analysis and its importance can be perceived since members of this group of toxins are potentially important candidates for mosquito control programs. However, the availability of an experimental structure for a complete active fragment of Cry10Aa, Cyt1A is expected to accelerate the development of engineered Cry and Cyt toxins that have increased potency and prolonged stability. This structure should be useful to combat insect resistance to the existing Cry toxins. It also helps to further stimulate the mechanistic studies aimed at dissecting the molecular steps for pore formation. About 70% of the currently identified *B.t* toxins are not active on insects, due to various reasons like low solubility in the insect gut environment, lack of binding to the BBMV

(Brush Border Membrane Vesicles) in the larval midgut and presence of protease cleavage sites. Hence, knowledge of the delta-endotoxins structure can be utilized to make these inactive toxins active by protein engineering.

ACKNOWLEDGEMENT

We are grateful for financial support from the DBT-BIF for providing us the infrastructure for performing this work.

REFERENCES

Alexandrov, N. N and Go, N. 1994. *Protein Science*. 3: 866-875.

Al-yahyaee, S.A. and Ellar, D.J. 1996. *Bioconjug Chem.* 7(4):451-60.

Aronson, A. I. , Geng, C. and Wu, I. 1999. *Appl. Environ Microbial.* 65: 2503-2507.

Balaraman, K.J. 2005. *Vect Borne, Dis.* 42: 81-86.

Bashford, D. 2004. *Front Biosci.* 9:1082-1099.

Ben-Dov, E., Zaritsky, A., Dahan, E. , Barak, Z. , Sinai, R. , Manasherob, R., Khameraev, A., Troyetskaya, A. , Dubitsky, A., Berezina, N. and Margalith. Y. 1997. *Appl. Environ. Microbiol.* 63:4883-4890.

Boonserm, P., Davis, P. and Ellar, D.J. and Li, J. 2005. *J Mol Biol.*, 348(2):363-82.

Boonserm, P., Mo, M., Angsuthanasombat, C. and Lescar, J. 2006. *J. Bacteriol.* 188(9): 3391-401.

Bosch, D., Schipper, B., van der Kleij, H., de Maagd R.A. and Stiekema, J. 1994. *Biotechnology.* 12:915–918.

Bravo, A., Gill S.S. and Soberón, M. 2005. Elsevier BV pp. 175–206.

Bravo, A., Sarabia, S., Lopez, L., Ontiveros, H., Abarca, C., Ortiz, A., Ortiz, M., Lina, L., Villalobos, F. J. , Peña, G., Nuñez-Valdez, M.E. , Soberón, M. and Quintero. R. 1998. *Appl. Environ. Microbiol.* 64:4965-4972

Bravo, A. 1997. *J Bacteriol.* 179:2793-2801.

Casida, J.E. and Quistad, G.B. 1998. *Ann. Rev. Entomol.* 43:1-16.

Chak, K.F., Chow, C.D. , Tseng, M.Y., Kao, S.S., Tuan S.J. and Feng T.Y. 1994. *Appl. Environ. Microbiol.* 60:2415-2420

de Maagd R.A., Bravo, A. and Crickmore, N. 2001. *Trends in Genet.* 17:193–199.

de Maagd RA, Weemen-Hendriks M, Stiekema W, Bosch D. 2000. *Appl Environ Microbiol.* 66:1559-1563.

Ferré, J., Real, M.D, Van Rie, J., Jansens, S. and Peferoen, M. 1991. *Proc. Natl. Acad. Sci.* 88:5119-5123.

Galitsky, N., Cody, V., Wojtczak, A., Ghosh, D., Luft, J. R., Pangborn, W. and English, L. 2001. *Acta Cryst D* 57: 1101-1109.

Gayathri, V., Jeyalakshmi, T. Shanmugasundaram, R. and Murthy, P.B. 2004. *J Environ Biol.* 25(4):419-421.

Glare, T.R., O'Callagan, M.2000. John Wiley and Sons,New York,350 pp.

Grochulski, P., Masson, L., Borisova, S., Pusztai-Carey, M., Schwartz, J.L. Brousseu, R. and Cygler, M. 1995. *Journal of Molecular Biology*, 254: 447-464.

Höfte, H. and Whiteley, H.R. 1989. *Microbiol Rev.* 53(2):242-255.

Hossain, M.A., Ahmed, S. and Hoque,S. 1997. *J. Invertebr. Pathol.* 70: 221-225.

Johnson, C. and Bishop, A. H. 1996. *FEMS Microbiol. Lett.* 142:173-177.

Kleywegt, G.J. and Jones, T.A. 1996. *Structure.* 4(12):1395-1400.

Kyte, J. and Doolittle, R.F.J. 1982. *Mol Biol.* 157(1):105-132.

Li, X.Q., Wei, J.Z., Tan, A. and Aroian, R.V. 2007. *Plant Biotechnol. J.,* 5:455–464

Li, Y., Wu, K., Zhang, Y. and Yuan, G. 2007. *Environ Entomol.* 36(5):1275-82.

Li, J., Koni, P.A. and Ellar, D.J.1996. *J Mol Biol.* 257(1):129-52.

Li,J., Carroll, J. and Ellar, D.J. 1991 *Nature* 353: 815-821.

Liang, Y., and Dean, D.H. 1994. *Proc. Nat. Acad. Sci.* USA 85: 7844-7848.

Martínez, C. and Caballero, P. J. 2002. Appl. *Microbiol.* 92:745-752.

Masson, L., M. Erlandson, M. Puzstai-Carey, R. Brousseau, V. Juárez-Pérez, and R. Frutos. 1998. *Appl. Environ. Microbiol.* 64:4782-4788

Morse, R. J. , Stroud, R. M., Yamamoto, T. 2001. Structure 9:409-417.

Ohba, M., Mizuki·, E. and Uemori, A. 2009. *Int J. Cancer Research and Treatment.* 29:1427-433.

Ohba, M., Wasano, N. and Mizuki, E. 2000. *Microbiol. Res.,* 155:17-22.

Parker, M.W. , Pattus, F. , Tucker, A. D. and Tsernoglou, D. 1989. *Nature,* 337:93-96.

Peng, D., Chen, S., Ruan, L., Li, L., Yu, Z. and Sun, M. 2007. *Food and Chemical Toxicology,* 45:1179-1185.

Pontius, J., Richelle, J. and Wodak, S.J. 1996. *J Mol Biol.,* 264(1):121-136.

Prabagaran, S.R., Jonathan Nimal, S. and Jayachandran, S. 2002. *Appl. Microbiol. and Biotech.* 102: 213-226.

Promdonkoy, B., Pathaichindachote, W., Krittanai, C., Audtho, M., Chewawiwat,N. and Panyim, S. 2004. *Biochem.Biophys. Res. Commun.*317:744-748.

Rausell, C., García-Robles, I., Sánchez, J. , Muñoz-Garay, C., Martínez-Ramírez, A. C.,Real, M. D. and Bravo, A. 2004. *Biochim Biophys Acta* 1660: 99-105.

Sánchez, R., Pieper, U., Melo, F., Eswar, N., Martí-Renom, M.A., Madhusudhan, M.S., Mirkoviæ, N. and Sali, A. 2000. *Nat Struct Biol.* 7 :986-990.

Sanger, F., Nicklen, S. and Coulson, A.R. 1977. *Proc. Natl. Acad. Sci. U.S.A.* 74(12): 5463-5467.

Schnepf, H.E., Cricmore, N., Vanrie, J., Lereclus, D., Baum, J., Feitelson, J., Zfider, D.R. and Dean. D.H. 1998. *Microbiol and Molecular Bioly Rev.* 62:775-806.

Smith, G.P. and Ellar, D.J. 1994. *Biochem J.,* 302: 611-616.

Thorne, L. , Garduno, F. , Thompson, T., Decker, D., Zounes, M., Wild,M., Walfield, A. M. and Pollock, T. J. 1986. *J. Bacteriology,* 166: 801-811.

Van Rie, J.1991. *Trends Biotechnol.*9:177-179.

Ward, E,S.and Ellar, D.J. 1986. *J. Mol Biol.* 191(1):1-11.

29

Molluscan Resources and Its Management Measures in Andaman & Nicobar Islands

S.N. Sethi, M. Balakrishnan, S. Chand and B.Varghese

INTRODUCTION

Since the onset of human civilization molluscs have a tremendous impact on Indian tradition and economy. In India, molluscs have been in great demand as ornaments, currency, as popular panacea for illnesses and as mascot for ward off evil spirits in olden days. Now, molluscs have assumedgreater significance in our industrial, technological and aesthetic aspects of life, in addition to edible utility. Several species of gastropods and bivalves are traditionally fished for food and shell from time immemorial. About 80,000 - 1, 00,000 species of molluscs are recorded from the world over and a total of 3271 species are recorded from India. They are represented in 220 families and 591 general and the spectrum comprise of 1900 gastropods,1100 bivalves,210 cephalopods,41 polyplacophores and 20 scaphopods (edible and ornamental) and 14 species of cephalopods are exploited on commercial basis in India(Appukkuttan *et al.*,1989). It is difficult to categories shells satisfactorily according to their uses as many are collected for several purposes. The main categories are shown in the Table-1.

The commercial important bivalves along the Indian coast are the clams, mussels, edible oysters and pearl oyster. The bivalves were exploited for shell, meat, industrial purpose and for the pearl. The edible bivalves and ornamental shells become more popular. Among bivalves, edible oysters, pearl oysters, clams, mussels, pen shells, scallops and cockles are important found in Andaman & Nicobar Islands.

EDIBLE OYSTERS

Out of the seven species of edible oysters reported from India. Andaman & Nicobar Islands have four species of edible oysters among them, *Crassostrea rivulari* and *C.gryphoides,* are commercially important species for mariculture but *C.madrasensis* is the mainstay of oyster fisheries of India but which is rare in this islands and *Saccostrea cucculata* is small but widely distributed along the Islands.

PEARL OYSTERS

There are five species of pearl oyster available in this islands and among them blacklip pearl oyster, *Pinctada margaritifera* is predominant. This pearl is unique in this island and widely distributed in this islands.

CLAMS

Among the exploited bivalve molluscans resources of India clams are widely distributed and abundant. They form subsistence fisheries along Indian the coast and fished by men, women and children from the intertidal region to about 4 meter depth. They are handpicked or collected by hand dredges. The commercially exploited clams are Vellorita cyprinoides, Meritrix meritrix, Katelysia opima, Anadara granosa. File clams, Cockles, Cardium sp., Trachycardium sp; Surf clams, Mactra sp.; Wedge clams, Razor clams, Macome sp., Tellina sp.; and Bean clams, Donax sp. are available in Andaman's. Giant clams, Tridacna crocea is most abundant species in this Islands, Tridacna maxima has commercial value. The other two giant clams Tridacna squamosa and Hippopus hippopus are rare. Plenty of blood clams, Anadara granosa are available in this Bay Island which has lots of export demands but reports are scanty on its distribution, density and biology aspects till now.

MUSSELS

Along the Indian coast two species of mussel's viz., the green mussels, *Perna viridis* and the brown mussels, *Perna indica* are commercially important. The green mussels exist in scattered in the intertidal regions and limited populations in Andaman's but it grows very large size and suitable for mass propagation to cater the needs of seafood eaters.

GASTROPODS

Important gastropods which occur exclusively in Andaman waters, but rare in mainland coast are top shell, *Trochus niloticus* , turban shell, *Turbo marmoratus*., king shell, *Cassis comuta* and queen shell, *Cyprecassis rufa*. These ornamental molluscs fetch lucrative prices. The top shell and turban shell fishery is banned since both the species are coming under scheduled one and schedule IV of wildlife (protection Act) 1972. The cowries which are abundant and have more than 35 species known from these islands. Among them, tiger cowries, *Cyprea tigris* is important gastropod. There are more than 50 species of cone shells belongs to conidae are known from these islands. The muricid gastropods comprise 22 species and important among them belong to *Drupa sp., Thasis sp., Murex sp.* Etc There are 5 species of abalones, 20 species of limpets, 26 species of top shells(), 10 species of turban shells (), 14 species of neritids, 17 species of periwinkles, 3 species of sun dials, 10 species of cerithids, 34 species of true conchs, 6 species of helmet shells, 6 species of tritons, 2 species of frog shells, 3 species of coral shells, 9 species of of dove shells, 2 species of whelks,30 species of nassa, 9 species of tulips, 11 species of olive shells , more than 60 species of mitres, 3 species of chanks, 3 species of harps, 5 species of nutmegs, 11 species of turrids, 24 species of augur shells, and scores of other species of non- descriptive gastropods occurring in these islands.

The chank or 'Indian conch', *Turbinella pyrum* Linnaeus, is a large, thick-shelled gastropod abundant in the shallow waters of Andaman. The Sacred Chank,*Turbinella pyrum* is an important gastropod exploited commercially in a few maritime states in India; the major portion is from Gulf of Mannar , Palk Bay and Andaman sea. The shell is sometimes known as the Sacred Chank because of its importance in both the Hindu and Buddhist religions, where it is used as a ceremonial trumpet and as a libation vessel (*jhal shankha*). The rare 'valampuri' chanks are especially venerated; in these the shells are abnormally coiled anticlockwise. They are usually donated to temples by wealthy patrons, often richly ornamented with gold or silver. The shells of sacred chank is extensively used in the bangle industry in west Bengal and exploited from time immemorial.

The edible gastropods, *Babylonia spirata&Babylonia zeylanica* are commonly known as 'Whelk,' or 'Spiral Babylon' and 'Puramuttai chank' (Dove egg shell) in local parlance and 'Baigai' in the trade sector, is a marine edible gastropod belongs to family Buccinidae. It is widely distributed in the Indo Pacific region. It is widely fished from Tamil Nadu at places such as Gulf of Mannar, Poompuhar, Nagapattinam, Puducherry and Andaman waters also.

Cephalopods

These cephalopods include squids, cuttle fish, octopus and nautilus. *Nautilus pompilius* is an important mollusk with high value in shell craft industry. The cuttle fishes available are *Sepia pharaonis* , *Sepia aculeate* , *Sepiella inermis*, S*ipioteuthis lessoniana* and squids *Loligo duvaucelii* , *L. singhalensis* and Octopus, *Octopus dollfusi* , *Cistopus indicus* and *O. membranceus* .

Nudibranchs

A lesser-known molluscan group is the nudibranchs, commonly called "sea slugs," which display a fascinating variety of colors and body forms. A nudibranch feature that is unique to molluscs is the radula, which is used to eat sponges, corals, anemones, hydroids, bryozoans, tunicates, algae, and sometimes other nudibranchs. Nudibranch (pronounced nu-da-brank) means "naked gill," an apt name because their gills are external. Some nudibranchs have gills towards the rear. Others have rows of respiratory projections called cerata situated along the body. On the front end of the nudibranch is a pair of rhinophores (rhino = nose, phore = carrier). These structures are sensory organs that detect chemicals in the water, similar to a sense of smell. Both the gills and the rhinophores can be retracted into specialized pockets for protection. Nudibranchs belonging to the subclass Opisthobranchia are among the least studied molluscs in India.

Sea slugs are an important group of molluscs, although they are not consumed as a food in India but many anti-cancer, anti-tumor and viral compounds are available could be the potential drugs from sea slugs. The sea slugs of the families Dorididae, polybranchidae are mostly found in coral reef areas of Andaman & Nicobar Islands.

Ornamental Molluscs

Several ornamental gastropods and bivalves with trade value are distributed in the Andaman and Nicobar Islands. The important shells are *Turbinella pyrum, Babylonia spirata* , *Babylonia zeylanica, Chicoreus ramasos, Cyprea sp., Conus sp., Trochus niloticus* (endangered), *Turbo marmoratus, Cymatium sp. ,Cymbium melo ,Drupa sp. ,Fistularia sp., Hemifusus sp., Lambis sp.*(endangered), *Murex murex, Nautica dydima, Nerita sp., Oliva sp., Strombus sp., Pyrene sp., Tonna dolium, Tibia sp., Dentalium sp., Umbonium sp.* They areregularly collected, cleaned and marketed and form the basic material for the shell craft industry. Majority of ornamental molluscs are landed as by-catch of the trawlers and used in the shell trade. Approximately 70 species are used in ornamental shell trade. Again approximately 70 species of gastropods are used in the trade at present. The rate gastropod collected includes: *Conus milneedwardsii* (endangered), *Conus bengalensis, Conus miles, Conus striatus* and *Conus geographus* from the family Conidae; *Strombus listeri* and *Strombus plicatus sibbaldi* (both endangered) from the genus Strombus and *Lambis crocea, Lambis truncate* and *Lambis scorpius* (three endangered) of the family Strombidae, *Cypracassis rufa, Charonis tritonis, Trochus niloticus* and *Turbo marmoratus* (all endangered). The giant clams, *Tridacna sp.* from Andaman & Nicobar Islands are endangered.

CONSERVATION AND MANAGEMENT STRATEGIES

In the present context molluscs appears to be least endangered or mostly exploited at optimum level. Commercial exploitation especially trawling account for higher fishing pressure leading to greater reduction of molluscan population in time wild in mainland , Other sources of depletion are pollution and environmental hazards especially in the intertidal areas in mainland and Islands. Clams are sessile animals prone to over exploitation, strict vigilance on exploitation of undersized clams from natural beds. Establishment of clam sanctuaries or Clam Park is suggested for replenishment of stock.

Edible oyster, green mussel fishery is of sustenance nature and there is no possibility of over exploitation at present. Farming activities of oyster may be initiated in large scale which having very bright scope in this islands. Giant clams *Tridacna sp.* is endangered animals and is protected under wildlife (protection) Act 1972. Sea ranching of giant clams may be initiated by setting hatcheries in these Islands.

The licensing system for fishing of chanks by conventional divers exists in Tamilnadu and Kerala. This system can be followed in this island for optimum exploitation of molluscs resources. We may have to think about quota system for many of these species especially *Conus sp., Strombus sp.,Cowries* sp., *Chicoreus sp., Mures sp.,* etc for conserving these resources.

Regulation of mesh size of the hand dredge and dredging mainly in estuarine and shallow coastal habitat – Kalanadi estuary & Vembanad Lake, Kerala – *Paphia malabarica, Meritrix casta* and *Villorita cyprinoides* practice may be applied for this Bay Islands for clams and bivalves resources. To replenish the clam resources, "Clam sanctuaries" or" Clam Park" in known clam fishing estuaries or bays area could be declared.

Issuing license for fishing (like chank) & Lime shell fishing rights issue of license as practiced by the State Dept. of Mining & Ecology (Kerala Miner Minerals Concession Rules, 1967). Imposing ban on fishing in the spawning season ("Fishing Holidays") – undersized chanks and Egg masses in Trawl catches along TN coast, similar license could be adapted for ornamental shells in this island.

Declaration as Prohibited areas for fishing : In Andaman & Nicobar islands – *Trochus sp.* & *Turbo* sp. – overexploitation by Divers lead to a decrease in the landings in recent years. So areas around Little Andaman, Nicobar, Katchal and Comorta island s up to 500m from shore line.Educating the fisher folk and forest department staffs about the need for conservation is very much essentials in this Bay Islands.

REFERENCES

Appukkuttan, K.K., Chellam.A., Ramadoss, K., Victor, A.C.C. and Meiyappan, M.M., 1989. Dynamics of Indian chank fisheries. *CMFRI Bull.*, 42(1):100-104.

FSC
www.fsc.org
MIX
Papier | Fördert
gute Waldnutzung
FSC® C083411

Zeitfracht Medien GmbH
Ferdinand-Jühlke-Straße 7
99095 Erfurt, Deutschland
produktsicherheit@kolibri360.de